The IBNET
Water Supply
and Sanitation
Performance
Blue Book

The IBNET Water Supply and Sanitation Performance Blue Book

THE INTERNATIONAL BENCHMARKING NETWORK FOR WATER AND SANITATION UTILITIES DATABOOK

Caroline van den Berg

and Alexander Danilenko

THE WORLD BANK
Washington, D.C.

ISBN: 978-0-8213-8582-1
eISBN: 978-0-8213-8588-3
DOI: 10.1596/978-0-8213-8582-1

Library of Congress Cataloging-in-Publication Data has been requested.

Cover photo: Alexander Danilenko
Cover design: Naylor Design

Contents

TABLES

Foreword

The provision of safe and reliable water and sanitation is a cornerstone of municipal services. Yet even as the demand for more and better services grows, the level of financing for these services becomes increasingly constrained. Thus, utilities around the world look ever more urgently for ways to improve their performance and provide better services at the lowest possible cost. One effective means for accomplishing this is by comparing their performance with that of similar utilities elsewhere. As a result, water and wastewater utilities require a source of comprehensive, reliable data as a basis for meeting their constituents' demands for high-quality services.

The International Benchmarking Network for Water and Sanitation Utilities (IBNET) helps to build the resources for meeting this demand and suggests ways of providing improved services. Funded by the U.K. Department for International Development (DFID) and jointly administered by the Water and Sanitation Program and the Water Anchor of the World Bank, IBNET provides the first global benchmarking standard for assessment of the water sector. Its database provides operational, financial, and technical indicators on more than 3,000 utilities in 100 countries that provide services for more than one-quarter of the world's urban population. Through its performance-assessment standards and continually updated database, IBNET serves as a global yardstick with which utilities and national policy makers, as well as the public, governments, municipalities, utilities, investors, and other users, can compare and evaluate the performance of water and wastewater utilities throughout the world.

Better understanding is the first step toward enhanced performance. This volume is designed to raise awareness of how IBNET and its tools can help governments to regulate their utilities more effectively and help utilities to improve their services. IBNET tools can also be used for process benchmarking, the normative comparison by one utility of their processes' and procedures' effectiveness against that of selected peers. Process benchmarking is particularly effective in a twinning arrangement involving the formal exchange of ideas and methods between two "sister" utilities; such comparisons, for example, of billing and collection systems, will reveal which system performs better. The more effective system can then be adopted by the underperforming utility.

Since its inception in 1997, IBNET has created partnerships with international donors, water utility associations, and regulators as well as with individual utilities and municipalities throughout the world to expand use of its database and to further strengthen benchmarking practice in the field. IBNET has played a key role in international reporting on the water sector. Since 2004, information collected by IBNET has served as the basis of more than 150 papers and reports on water sector status, performance, and economics. Such reporting builds understanding and brings transparency into the sector as well as helping to improve water services for all, including the poor. Most of the utilities that regularly collect and report their performance information to IBNET have consistently improved their financial and technical performance.

As a tool available to donors and developing agencies, IBNET helps to address water sector issues in poor and developing countries. It is widely used to justify the Bank's strategic involvement in the sector and to monitor sector development: about 10 World Bank projects have used IBNET during project preparation and in support of proposed investment programs. In a few countries, IBNET already serves as a foundation for sector strategy and investment planning. During development of these programs, planners have relied on the fact-based, objective information provided by IBNET and its tools.

Global initiatives such as the new Hashimoto Action Plan and DFID Water Action Plan call for increased monitoring and reporting at the global and national levels. IBNET's benchmarking successfully harmonizes existing monitoring and reporting activities in the water-supply and sanitation sectors to improve utilities' service delivery.

The IBNET Blue Book creates a baseline and, at the same time, offers a global vision of the state of the sector in developing countries. By tracking progress in and quantifying and assessing the water supply and sanitation sectors, IBNET helps meet the goal of providing safe, sustainable, and affordable water and sanitation for all. We invite water and sanitation service providers, municipalities, government authorities, and all users of water services to join us in this effort.

Jae So
Water and Sanitation Program Manager
The World Bank

Julia Bucknall
Water Anchor Manager
The World Bank

Acknowledgments

This report, a joint effort of the Water and Sanitation Program (WSP) and Water Anchor of the World Bank, was prepared by a World Bank team led by Caroline van den Berg and Alexander Danilenko, and supported by John Bachmann of AECOM International Development. WRc plc developed and tested the original set of water benchmarking tools. A number of Bank staff members provided guidance and contributions at various stages, including Vivien Foster, Joseph Gadek, William Kingdom, Alain Lucassol, Philippe Marin, Abel Mejia, Josses Mugabi, and Dennis Mwanza. Special thanks to all WSP staff for their help in collecting, analyzing, and presenting the information contained here. The inputs from Masroor Ahmad, Mohammad Akhtaruzzaman, Vandana Bhatnagar, Wambui Gichuri, Abdul Motaleb, Hang Diem Nguyen, Lilian Otiego, Farhan Sami, Almud Weitz, and many other WSP and World Bank staff were highly valuable. We also thank Caroline Simmonds for her sector insights and professional editing.

Our great appreciation goes to the U.K. Department for International Development (DFID), which continues to fund and support the major part of the IBNET program.

IBNET Partners

ADERASA, Association of Water and Sanitation Regulatory Entities of the Americas
AfWOP, Africa Water Operators Partnership
EAP Task Force, OECD
SEAWUN, South East Asia Water Utilities Network

Albania: National Regulator of Water and Canalization (DRUK)
Belarus: Belcommunproject, Design Institute
Brazil: National System of Information of Water and Sanitation of the Ministry of Urbanization, SNIS
China: Shandong Provincial Water Association, SWA
Croatia: National Association Hrvatske Vode
Czech Republic: Institute for Structural Policy, IREAS
Former Yugoslav Republic of Macedonia: National Association of Water and Sanitation Utilities
Georgia: National Association of Water Utilities, Georgia Tskhalkanali
Hungary: National Environment Research Center at the Corvinus University, REKK
Kazakhstan: National Association of Water Utilities, Arna Su
Kyrgyz Republic: National Association of Communal Services Providers, Kyrgyzzhilcommunsoyuz
Moldova: Moldova National Association of Water and Wastewater Utilities, AMAC
Philippines: Philippine Water and Wastewater Association, PWWA
Romania: National Association of the Potable Water Utilities, ARA
Russian Federation: Moscow Institute for Urban Economics, IUE
Ukraine: Institute for Municipal Development, MDI
Vietnam: Vietnam Water Association, VWA

Abbreviations

ADERASA	Association of Water and Sanitation Regulatory Entities of the Americas
AMAC	Moldova Apa Canal
conn	connection
CRR	cost recovery ratio
DFID	U.K. Department for International Development
EU	European Union
GNI	gross national income
IBNET	International Benchmarking Network for Water and Sanitation Utilities
I & I	inflow and infiltration
IWA	International Water Association
KPI	key performance indicators
lcd	liters per capita per day
m^3	cubic meter
MDG	Millennium Development Goals
NRW	nonrevenue water
OCCR	operating cost coverage ratio
OECD	Organisation for Economic Co-operation and Development
O&M	operation and maintenance
PPP	purchasing power parity
SNIS	National Sanitation Information System, Brazil
W	water
WSP	Water and Sanitation Program
WW	wastewater
WWTP	wastewater treatment plant

1

IBNET: THE INTERNATIONAL BENCHMARKING NETWORK FOR WATER AND SANITATION UTILITIES

Introduction

Water—essential to sustain life and livelihoods—is a core sector of the global economy. The water and wastewater utilities of developing countries generate a substantial portion of the sector's estimated annual turnover of US$500 billion (Global Water Intelligence 2009). In urban areas, these utilities play a key role in efforts to attain the Millennium Development Goals (MDGs) of halving by 2015 the population lacking access to improved water and sanitation. Enhancing the operational and financial performance of these utilities will provide the basis necessary for expanding access and improving quality of service.

The need for improved performance is not limited to developing countries. Urban water and wastewater utilities are under increasing pressure to perform. Among the many problems they face are volatile energy prices, a threat to the financial viability of their operations; in many countries, a short supply of capital improvement loans; and the uncertainties of climate change. In addition, regulators and citizens demand increasingly higher standards of environmental, social, and economic sustainability.

If water and wastewater utilities are to meet these increasing demands and expectations in both developed and developing countries, they must first take stock of their performance over time. Comparisons with similar utilities elsewhere in the country or region or with standards of international good practice can shed light on how well a utility is performing, identify areas for improvement, and help indicate a plan of action. A major challenge for measuring, and eventually benchmarking, water and wastewater utility performance has been the lack of standardized information. In only a few cases has a standard set of indicators been applied consistently to measure utilities' financial and operational performance.

The International Benchmarking Network for Water and Sanitation Utilities (IBNET), launched in 1996, provides options for standardized measurement of utilities' operational and financial performance. IBNET has established the first global benchmarking standard for water and wastewater utilities, providing a global yardstick against which utilities and policy makers can measure their performance and thus gain a better understanding of their strengths and weaknesses. Building on the achievements of other utility benchmarking efforts, notably those of the International Water Association, IBNET, administered under the Water and Sanitation Program of the World Bank and financed by the United Kingdom's Department for International Development (DFID), has grown from a short-term initiative to the largest publicly available water sector performance mechanism for collecting utility performance data. IBNET provides the tools to analyze these data and provides access to information on the performance of more than 2,500 water and wastewater services providers from 110 countries (although not all utilities report consistently). IBNET's four-language, Internet-based interface registers 3,000 users who download up to 10,000 benchmarking reports a month; IBNET information is widely used by utilities, researchers, consultants, investors, and donors.

This report serves three purposes. First, it aims to raise awareness of how IBNET can help utilities identify ways to improve urban water and wastewater services. Second, it provides an introduction to benchmarking and to IBNET's objectives, scope, focus, and some recent achievements. Third, it elaborates the methodology and data behind IBNET and presents an overview of IBNET results and country data.

By providing comparative information on utilities' costs and performance, IBNET and this study can be used by a wide range of stakeholders, including

- *Utilities:* to identify areas of improvement and set realistic targets
- *Governments:* to monitor and adjust sector policies and programs
- *Regulators:* to ensure that adequate incentives are provided for improved utility performance and that consumers obtain value services
- *Consumers and civil society:* to express valid concerns
- *International agencies and advisers:* to perform an evaluation of utilities for lending purposes
- *Private investors:* to identify opportunities and viable markets for investments

What Is Benchmarking?

Scrutiny of business practices has intensified in recent years, and the need for transparent and standardized information with which to compare utilities' performances has gained prominence, leading to increased emphasis on measurement of results, on transparency, and on accountability. As a result, the use of benchmarking has increased, and its value is widely recognized. The primary objectives of benchmarking are as follows:

1. To provide a set of Key Performance Indicators (KPIs) related to a utility's managerial, financial, operational, and regulatory activities that can be used to measure internal performance and provide managerial guidance

2. To enable an organization to compare its performance on KPIs with those of other relevant utilities to identify areas needing improvement, with the expectation of developing more efficient or effective methods to formulate and attain company goals as set forth in its business plan

Two types of benchmarking can be distinguished. *Metric benchmarking* involves systematically comparing the performance of one utility with that of other similar utilities, and even more importantly, tracking one utility's performance over time. A water or wastewater utility can compare itself to other utilities of a similar size in the same country or in other countries. Similarly, a nation's regulators can compare the performance of the utilities operating there. Metric benchmarking, essentially an analytical tool, can help utilities better understand their performance. Such benchmarking is most powerful when carried out over time, tracking year-to-year changes in performance.

Process benchmarking is a normative tool with which one utility can compare the effectiveness of its processes and procedures for carrying out different functions to those of selected peers. A utility can compare its billing and collection system, for example, to those used by other utilities to see which system performs better. When the comparison reveals one utility's system to be more effective or efficient than the other's, the underperforming utility can adopt and internalize those processes and procedures as appropriate. The performance indicator constitutes the building block of both types of benchmarking. Indicators are quantitative, comparable measurements of a specific type of activity or output. Often based on ratios and percentages, water sector indicators measure, for instance, the percentage of population served by the piped water-supply network or a utility's ratio of total revenues to total costs during a given year.

What Is IBNET?

IBNET provides a set of tools that allows water and sanitation utilities to measure their performance both against their own past performance and against the performance of similar utilities at the national, regional, and global levels.

The IBNET Toolkit

IBNET consists of three major tools. The first is the IBNET Data Collection Toolkit, which can be downloaded from the IBNET Web site at http://www.ib-net.org; this Excel spreadsheet indicates a set of data to be completed and offers detailed instructions on the precise data to enter. The second tool is a continuously updated database of water and sewerage utilities' performance. This database allows utilities and other sector stakeholders to search for data in different formats and provides the means for simple benchmarking of utility data. The benchmarking tool enables the utility to compare itself to other utilities with similar characteristics (for example, size, factors related to location, and management structure). The third tool provides data on participating agencies. This information helps organizations interested in measuring utility performance to contact neighboring utilities and other organizations to build local networks for performance assessment and benchmarking.

IBNET's Key Organizational Aspects

IBNET has three key aspects. The first is that participation is voluntary, with the result that organizations contributing to IBNET are very diverse. They include, for example, regulatory associations (such as the Association of Water and Sanitation Regulatory Entities of the Americas [ADERASA]), national water associations, government departments and agencies involved in monitoring urban water supplies and sewerage utilities, and, more recently, individual utilities.

A second feature of IBNET is that it does not itself collect data. Rather, it sets up mechanisms by which many different organizations conduct data collection. From its start, IBNET's strategy has been to use a highly decentralized approach. Those closest to the utilities and most knowledgeable about local conditions are best suited to compile data and assess the utilities' performance. IBNET's role is to provide instruments, such as the IBNET Toolkit, to support this process. IBNET also organizes workshops to assist local agencies in training staff members in data collection and analysis, and it provides feedback once the data are collected. In its feedback, IBNET checks the quality of the data to ensure internal consistency and helps participants to analyze the data. Experience has shown that after the data collection process has been repeated several times, this technical assistance becomes increasingly redundant, and the organizations can thenceforth undertake data collection on their own.

The third key IBNET feature, one fairly rare among agencies involved in utility benchmarking, is its focus on developing time-series data. Without time-series data, trends in utility performance and the impact of water and sanitation policies are difficult to detect. Effective development of time-series data requires ensuring that the data remain comparable over time through the rigorous use of a standardized data set and indicators as well as frequent data updating. In IBNET practice, most of the data are updated every two years. As performance assessment and benchmarking gain more prominence in the sector as regulation and monitoring tools, obtaining data on an annual basis has become easier, especially in countries with increasingly institutionalized performance assessment. Currently, more than 50 percent of utilities in IBNET have at least 4 years of data results, and a large percentage of utilities represented in the IBNET database have data series extending between 5 and 10 years. This database allows innovative time-series performance analysis as well as cross-section analysis.

What Can IBNET Do for You?

IBNET is a broad and versatile tool that offers different benefits to different types of users (see table 1.1). For water and wastewater utilities, IBNET provides a ready-made analytical tool for self-assessment of performance at no cost to the user. By participating in IBNET, utilities can analyze their strengths and weaknesses in relation to those of peer organizations and can track their own performance over time. The results of the IBNET analysis can then be used to inform strategic business planning processes designed to improve management performance.

Both utilities and associations can exploit IBNET-based assessments to position themselves to receive financing for capital improvements. Where national policy makers are interested in making capital financing available, IBNET can be adopted as an analytical tool for assessing needs and allocating resources. Private investors interested in expanding their interests in the water and wastewater sector can also use IBNET to carry out an initial screening of potential target utilities. A broad-brush IBNET analysis will provide a reliable assessment of the strengths and weaknesses of different utilities, pinpointing those with revenue-generating potential using an analysis of financial results, service-delivery efficiency, and customer-relations management. The results of an IBNET assessment can be

Table 1.1   IBNET Benefits by Type of User

User	Benefits
Utilities and utility associations	• Self-assessment of performance
	• Justification for requests for financial and other assistance (facilitates borrowing money)
	• Focus on shortcomings, providing strategic business planning baseline
	• Analytical platform for process benchmarking through twinning arrangements
	• (*For associations*) Facilitation of utilities' participation through information exchange
	• (*For associations*) Provision of data to inform advocacy for the water and wastewater sector
Regulators	• Assessment of performance to underpin tariff setting
	• Comparative analysis of utilities' performance
National policy makers and international donors	• Evaluation of sector in relation to other cities, regions, or countries
	• Focus on shortcomings, providing strategic planning baseline
Private operators and investors	• Comparative analysis of utilities' performance
	• Focus on strengths and weaknesses, enabling due diligence
Researchers and consultants	• Comparative analysis of sector performance
	• Comparative analysis of a utility performance

Source: IBNET.

Box 1.1   Brazil: Formalizing Performance Assessment into Law

Brazil provides an example of how benchmarking can drive water or wastewater sector reform. Starting in 1992, the World Bank financed Brazil's Water Sector Modernization Program, establishing a national system for measuring the performance of water and wastewater utilities. The National Sanitation Information System (SNIS) began to collect information on service quality, financial performance, institutional efficiency, and other parameters. SNIS now has data on more than 600 utilities representing more than 4,000 municipalities. (Many utilities are regional in scope.) The recently approved national water law upgraded the performance-measurement system and made it the nerve center of a national performance-improvement initiative. Substantial funding under the Growth Acceleration Program has been earmarked for capital improvement in water and especially wastewater systems. Funding eligibility decisions are made on the basis of performance criteria calculated using the SNIS system. In effect, the focus on results-based management created the need to measure performance accurately and quantitatively. With the help of a performance-measurement system similar to IBNET, Brazil has launched its national water and wastewater sector on a transparent course toward improved management and better service delivery. Following its success with water and wastewater utilities, SNIS has expanded its benchmarking to companies providing solid-waste services.

Source: SNIS, Brazil.

---

used to write the terms of reference for the more detailed due diligence exercises required before final decisions on an investment are made.

The cases of Moldova and Brazil, detailed in boxes 1.1 and 1.2, show how IBNET can be used to refine and coordinate national water and wastewater service-improvement programs by introducing results-based management and systematic performance measurements for participating utilities. These

---

Box 1.2 Moldova: Using Performance Assessment for Advocacy

Moldova Apa Canal (AMAC), a nongovernmental association of water and wastewater service providers, in 2001 teamed up with the Organisation for Economic Co-operation and Development (OECD) to test the Water Performance Assessment Start-Up Toolkit, the predecessor to IBNET. Data were collected from participating utilities retroactively for the period 1996 through 2000. The data collection standard was modified in 2004 with IBNET's introduction in Moldova.

The IBNET data clearly showed that investment was required to replace deteriorated water and wastewater systems. AMAC recommended to the government that World Bank loan funds be used to finance replacement of piped networks and energy-inefficient equipment. The selection of utilities that would receive loan financing was carried out using IBNET indicators. More than US$20 million has been invested since 2001 in eight water- and wastewater-improvement projects across Moldova.

Source: Moldova Apa Canal, National Association of Water and Wastewater Companies.

---

countries' experiences with the method demonstrate how effective performance benchmarking can be in facilitating national or regional efforts to reform the water and wastewater sector. First, benchmarking provides a comprehensive, global view of the performance of a nation's utilities. Further, it correlates technical performance with financial performance and calculates some measures of the overall efficiency of an individual utility's operations. Only with such a broad perspective can policy makers reach informed decisions about the best direction in which to take the sector as a whole and how best to steer the sector toward stated goals and objectives.

IBNET Achievements

The water industry is a core sector of the economy. In 2007, Global Water Intelligence estimated the current market for urban water supply and sewerage handling to be US$210 billion in 2006, of which the market in developing countries accounts for US$80 billion. The rural market is significantly smaller, at US$15 billion, especially in view of the large populations living in these areas.

The IBNET database includes basic performance data for about 2,600 water utilities between 1995 and 2008. The database represents more than US$27 billion in annual revenues in 2006, that is, about 39 percent of the official water market and 32 percent of the total official and gray, or unofficial, water market in developing countries, as calculated by the Global Water Intelligence Unit (see table 1.2). (As IBNET is especially active in middle-income countries, it is likely that the Global Water Intelligence figures may underestimate the real size of the developing countries' water markets.) For 2008, in terms of these countries' total population of urban households with piped-water access, IBNET covered 256 million water-supply users and 157 million users of sewerage or sanitation from a total of about 1.7 billion people. That number represents approximately 15 percent of the population, a calculation based on the UNICEF-WHO Joint Monitoring Program 2008 MDG assessment at http://www.wssinfo.org. (IBNET's data collection process has not been finalized; it is still on-going in several parts of the world, so these data may show changes over time.)

Table 1.2   IBNET Representation as Percentage of Estimated Total Urban Market Size in Developing Countries

Region	Estimated urban market size in developing countries in US$ billion			Estimated operating revenues in IBNET as % of urban market share	
	Official	Gray	Total	Official	Total
Africa	3.8	2.0	5.8	44	29
East Asia and Pacific	27.8	4.5	32.2	20	18
Europe and Central Asia	16.0	2.4	18.4	37	32
Latin America and Caribbean	15.2	3.5	18.7	82	66
Middle East and North Africa	1.6	0.6	2.2	8	6
South Asia	1.1	1.7	2.8	15	6
Total developing countries	65.9	14.7	80.6	39	32

Source: Global Water Intelligence, *Global Water Markets 2007*, IBNET.

Since its inception, IBNET can lay claim to a number of achievements in the water and wastewater sector. Foremost has been its role as the first global benchmarking standard for the sector. Other accomplishments include the following:

- IBNET has contributed to improved knowledge and understanding of benchmarking, including awareness that performance can and should be measured in a comprehensive way, taking into account the utilities' financial, institutional, and technical dimensions.
- IBNET efforts have helped participating utilities to achieve more thorough understanding of their performance in relation to that of their peers and to improve their managers' strategic focus. Some of these managers have used their improved understanding to formulate plans for future improvement.
- Since its inception in the 1990s, IBNET has accumulated the largest public database on water and wastewater utilities and is thus able to provide utilities and others interested in the water and sanitation sector with performance data from nearly 3,000 utilities in 110 countries for the period from 1995 to 2010.
- About 63 percent of the utilities represented in the IBNET database have more than four entries regarding performance, making it increasingly possible to examine performance trends at the utility and sector levels.
- With funding from DFID, initiated in 2005, IBNET concluded technical assistance agreements with many organizations throughout the world. IBNET has since provided support to numerous organizations seeking to hone their performance assessment and benchmarking skills. The organizations include the national associations of Georgia, Moldova, Romania, the Former Republic of Macedonia, Bulgaria, Kazakhstan, Serbia, and Vietnam and the Shandong and Liaoning provincial water associations in China. In a number of countries, including Albania, Armenia, Belarus, the Russian Federation, Ukraine, the Kyrgyz Republic, Hungary, Poland, the Czech Republic, and Sudan, IBNET helped inaugurate benchmarking efforts. With the support of the Water and Sanitation Program–South Asia, IBNET benchmarking was recently begun in Bangladesh, India, and Pakistan.

Table 1.3   Number of Utilities in IBNET by Region

Year	Africa	East Asia and Pacific	Europe and Central Asia	Latin America (including United States and Canada)	Middle East and North Africa	South Asia	Total
1994	0	0	0	0	12	0	12
1995	4	22	23	0	12	1	62
1996	13	21	64	26	12	5	141
1997	13	83	148	26	12	0	282
1998	14	83	157	27	12	0	293
1999	16	83	157	27	0	0	283
2000	46	83	312	229	4	0	674
2001	45	93	760	267	0	7	1,172
2002	60	116	788	296	0	4	1,264
2003	62	155	841	601	0	4	1,663
2004	95	200	854	650	1	13	1,813
2005	75	148	427	503	1	24	1,178
2006	62	171	428	706	1	18	1,386
2007	50	190	389	605	0	11	1,245
2008	45	63	270	722	0	11	1,111

Source: IBNET.

- The number of data observations on the IBNET Web site has grown exponentially. Currently, the database contains almost 500,000 data observations, compared with 345 in 1997. These observations form the basis of a much larger set of performance indicators, available to the general public on the IBNET Web site, http://www.ib-net.org (see figure 1.1 and table 1.3).

- In 2010, IBNET published a tariff database providing data on water and wastewater tariffs in more than 210 utilities worldwide. The tariff database reports the water price charged to domestic users per cubic meter for the first 15 cubic meters consumed, delivered through a 20-millimeter (5/8-inch) pipe (see figure 1.2).

- IBNET plays a key role in international reporting on the status of the water sector. Since 2004, more than 150 papers and reports on water sector status, performance, and economics have been published based on indicators collected by IBNET.

Yet the ultimate value of utility benchmarking is the extent to which it leads to greater efficiency and delivery of better services. More than one country has made IBNET or similar performance measurement systems the core of its national efforts at utility reform. These efforts demonstrate that, where adopted, performance assessment and benchmarking improve performance. This result holds for all contexts, whether in low-, middle-, or high-income countries. Interestingly, not only does performance improve, but the variance in performance across utilities decreases: although the number of utilities in the database has increased rapidly over this period, performance as measured by the operating cost coverage ratio (measuring how many times operating revenues cover operation and maintenance costs) has remained stable—despite the triple impact of fuel, food, and financial crises (see figure 1.3).

Figure 1.1   IBNET Country Coverage

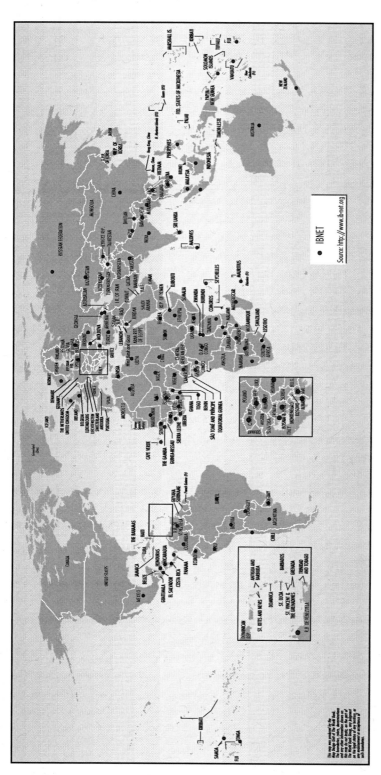

Source: IBNET.

Figure 1.2   IBNET Water Tariff Coverage

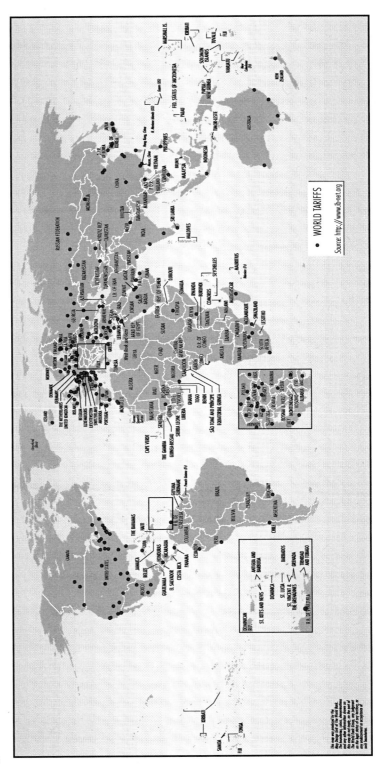

Source: IBNET.

Figure 1.3   Median Operating Cost Coverage Ratio

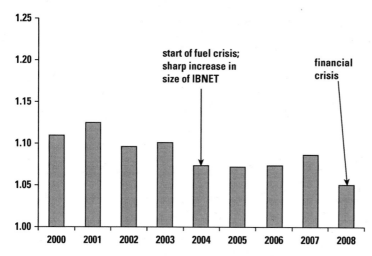

Source: IBNET.

Second, benchmarking promotes transparency. When the same data are collected from each utility, benchmarking allows direct comparisons between service providers with respect to operational results, system conditions, service quality and coverage, financial condition, customer affordability, and other dimensions of utility performance. For public companies, such reporting is often a statutory requirement, addressing customers' rights to see how their money is used. For private urban utilities or utilities intending to go private, publishing performance data represents both sound corporate governance and a way to attract private capital.

Third, performance benchmarking is an effective tool for rationalizing the use of scarce resources. When, for example, Utility A's water network reaches only half of the households in its jurisdiction, while Utility B's network reaches four-fifths of the households in its jurisdiction, clearly, all other things being equal, Utility A should be given priority in the distribution of public funds for network extension. But if Utility A has a high revenue collection backlog or a low level of operational cost recovery, then more information on its financial management capacity should be gathered before lending decisions are made.

As the previous example illustrates, benchmarking is most effective when combined with due diligence. By definition, a broad-brush picture of utility and sector performance, benchmarking is not intended to be the first and only source of input for decisions on investment, policy change, or changes in service level. Rather, benchmarking constitutes a cost-effective tool providing sector managers, including independent regulators, ministries, provincial governments, municipal authorities, and investors, with a bird's-eye view of the utilities' overall performance that can be used to prioritize needs and establish the main directions for new policies and programs. Ultimate funding decisions usually require probing more deeply using other tools, such as financial and technical audits by potential private investors or, in the case of the allocation of public monies for capital investment, due diligence on data submitted by the utilities.

2

IBNET METHODOLOGY

The International Benchmarking Network for Water and Sanitation Utilities (IBNET) data are collected at the local or national levels. Key sector institutions, such as water and wastewater associations, regulators, or research institutes working with these associations, typically reach out to their members to collect the baseline data needed to calculate indicators. The IBNET program often provides small-scale technical assistance to facilitate data collection.

Participants enter data into a standardized Excel spreadsheet under the categories General, Service Area, Water Service, Sewerage Service, Financial, and Tariffs. The spreadsheet can be downloaded easily from the IBNET Web site. (See appendix C for a list of the data items and indicators.) Macros in the spreadsheet automatically calculate the more than 27 groups of quantitative indicators that characterize the utility's performance with respect to water and wastewater coverage and quality, water consumption and production, cost recovery, operations, financial status, technical efficiency, billings and collections, and capital investment. Following completion of data entry and submission of the spreadsheet to the IBNET program, the World Bank's Water and Sanitation Program performs quality control on the data submitted and then enters the data into the IBNET database.

IBNET data can be accessed at no charge at http://www.ib-net.org. The interface allows users to create tables and graphs showing indicator values by utility, country, or region. The user can customize the tables and graphs to show only specified indicators, for example, the technical or financial performance of a given utility. From these, more complex tables can be constructed to show a number of utilities' performances on the same indicator. Results can be shown for a specific year or for a number of years. Finally, country reports (see figure 2.1) provide snapshots of national conditions across all utilities represented in the database.

For more targeted analysis, filters can be used to select utilities in specific countries or within specific population ranges or to select by indicator or year. Outputs appear in graphic format where time-series data are requested and available, and tables and charts can be copied and saved.

In addition to access to the database, the IBNET Web site provides methodological explanations and instructions on benchmarking and measuring water and wastewater performance. Step-by-step instructions guide users through benchmarking exercises. The site defines different methodologies, and bibliographies listing other methodological documents are provided. Example terms

Figure 2.1   Example of a User-Generated Country Report: Armenia

The International
Benchmarking Network
for Water and Sanitation
Utilities

IBNET   **Country Report**

Armenia

Indicator	2004	2005	2006	2007	2008
1.1 Water Coverage (%)	66	68	79	80	80
2.1 Sewerage Coverage (%)	51	46	35	34	35
4.1 Total Water Consumption (l/person/day)	156	126	153	146	151
4.7 Residential Consumption (l/person/day)	119	92	105	92	94
6.1 Non Revenue Water (%)	78	81	84	85	84
6.2 Non Revenue Water (m3/km/day)	107.5	129.1	109.5	108.8	94.7
8.1 % Sold that is Metered (%)	47	57	69	75	78
11.1 Operational Cost W&WW (US$/m3 water sold)	0.24	0.33	0.33	0.41	0.44
12.3 Staff W/1000 W pop served (W/1000 W pop served)	1.5	1.3	1.7	1.6	1.6
18.1 Average Revenue W&WW (US$/m3 water sold)	0.15	0.24	0.29	0.41	0.47
23.1 Collection Period (Days)	633	251	455	236	266
23.2 Collection Ratio (%)	66	64	72	83	87
24.1 Operating Cost Coverage (ratio)	0.64	0.71	0.88	1.00	1.05

Source: IBNET.

of reference make it easy for users to set up performance benchmarking at the national or regional level.

The IBNET site also facilitates networking within the benchmarking community by providing contact information for regional and national organizations active in benchmarking and performance measurement in the water and wastewater field.

IBNET's Limitations

IBNET works best as part of a comprehensive initiative to improve sector performance. The usefulness of benchmarking is seriously limited when utilities or other organizations neglect other appropriate steps. A simple peer comparison, for example, provides only a static view of performance. The proper approach to benchmarking involves three steps:

- Measure the real differences in performance among peers for key goals. This requires knowledge of the peer group adequate to ensure that the comparison is between "apples and apples."
- Investigate the reasons for the differences and develop strategies and tactics for improvements if organizations fall significantly below the best-practice standard drawn from analysis of the peer group.
- Implement definitive steps and programs to achieve needed improvements and carefully monitor the results. All projects of consequence should be monitored for performance to reveal what works and what doesn't.

Poor-quality data will also limit the usefulness of benchmarking. The quality of the IBNET database depends on the quality of the data submitted by individual utilities and utilities' associations. Some utilities submit precise, reliable data;

others do not. IBNET has tools and instruments (described in the section titled "IBNET Data Quality") with which it checks data quality, thus helping utilities to find obvious mistakes in their data submissions. Experience shows that, over time, utilities improve their skills in data collection and analysis. The differences in data quality resulting from this learning curve must be traded off against the benefit to the utilities of gaining the ability to measure results with accountability and transparency.

IBNET's data are further limited by the voluntary nature of membership. Some utilities are hesitant to submit their data. Only aggregated data are distributed or downloadable, however, which helps to make participation somewhat more attractive to these reluctant utilities. Publicly owned utilities have no objection to publishing data or, at least, indicators; these utilities are accountable to their governments and customers and, thus, as a matter of governance policy, must disclose basic technical and economic information about their operations.

IBNET participation is also largely limited to developing countries. While some Western European and Australian utilities contribute data, many others do not. Data are available for utilities in many developed countries, but with some exceptions no tradition exists even among publicly owned utilities of sharing this information. IBNET's global reach would be expanded considerably with the wider participation of European and North American utilities.

IBNET Data Quality

As noted above, the quality of the IBNET database depends on the quality of the data submitted by individual utilities and utilities' associations. IBNET therefore invests substantial effort in making sure the data are of the highest possible quality and accurately and adequately reflect the reporter's performance.

IBNET data come from a variety of sources, some of which have excellent quality assurance procedures (as in the case of regulatory data) and others of which follow less sound procedures. To correct for this, IBNET continually improves its data-checking procedures and makes users aware of the quality (or lack of quality) of particular data. The need for rigorous quality assurance procedures is always balanced against the need to avoid discouraging potentially valuable data sources from participating.

Data Quality at the Collection Level

The IBNET data collection tool contains ranges and built-in filters that prevent assembly of obviously wrong information. Among these mechanisms are, for example, that the population served by the utility cannot be more than 30 million, water production and consumption must be within reasonable levels, the volume of billed water cannot be higher than the volume produced, and the service provider's total revenue cannot be greater than the sum of its water and wastewater revenue. The toolkit thus allows the utility to review the consistency of its data immediately as they are collected. This helps prevent data fraud, as the system makes it substantially easier for the data collector to provide accurate data.

Every data collection report must be furnished to the IBNET team after the collection exercise and must provide both the sources of the data and the descriptions of their origin according to specific criteria for value and quality, as outlined in table 2.1.

Table 2.1 IBNET Value Categories for Data Quality

Value	Explanation of value
1	Based on sound records, procedures, investigations, or analyses that are properly documented and recognized as the best available
2	Derived generally as for the confidence rating, but with minor shortcomings; for example, some documentation may be missing, an assessment may be out of date, or some data may rely on unconfirmed reports or extrapolation
3	Extrapolated from a limited sample about which the collector is confident
4	Based on the best estimates of the utility staff members, without measurement or documented evidence

Source: Authors.

The data collector examines the calculated performance levels provided by all the utilities for sense and consistency, noting the following characteristics in particular:

- Data are within the ranges to be expected.
- Time trends appear to be reasonable.
- Confidence ratings assigned are as expected based on experience.

The data collector resolves any data quality concerns through discussion with the utility or water utility association and removes any data for which concerns cannot satisfactorily be resolved.

The IBNET Team Review

The IBNET team receives the data set and submits each datum to thorough review, focusing on outliers, data sources, and consistency. The team examines the calculated performance levels provided by all the utilities for sense and consistency to ensure that data are within the expected ranges and that time trends appear reasonable. By calculating averages for the given set of data, the team determines outlier utilities and reviews their performance jointly with the data collector.

Data Verification at the Uploading Stage

The IBNET team and its experts examine for sense and consistency the calculated performance levels provided at the country level. Once again, IBNET resolves any concerns over data quality through discussion with the data collectors and removes any data for which its concerns cannot be satisfactorily resolved.

Not all data are available during the first round of collection. In most cases, the financial data will be better collected and monitored than the technical performance data; these come from the utilities' technical departments and often are not readily available. In practice, however, during subsequent and follow-up data collection efforts these issues are usually resolved, and the processing and quality of all data tend to improve with each collection round.

3

STATUS OF THE SECTOR

Trends in Sector Status

Improving utilities' performance requires time and effort, the results of which can only be seen in the context of past performance. This makes trend analysis central to the improvement process.

Trend analysis is somewhat complicated, however, by some significant changes in the economic climate over the past few years. Energy makes up a significant part of the total operation and maintenance costs for many utilities, and between 2003 and 2007, fuel prices increased rapidly. In 2008, the median utility spent 23 percent of its total recurrent costs on energy. The fuel crisis and the consequent higher, more volatile fuel prices have affected many utilities, putting pressure on their ability to cover operation and maintenance costs with operating revenues. The fuel crisis was followed by a food crisis that was in turn followed in 2008 by a financial crisis, all further undermining utility revenues, as many customers suffered reverses and were unable to pay.

This chapter focuses on the trends in water and sewerage coverage in many countries, especially in the developing world, where efforts to widen access to a safe water supply and sanitation services have intensified as part of the larger effort to achieve the sector's Millennium Development Goals. These trends will demonstrate the progress that has been made in reforming the water utility sector. We will measure the performance of water utilities based on a set of indicators for operational efficiency, financial sustainability, and customer responsiveness.

Operational efficiency assesses the utility's use of inputs in the course of daily management. Operational efficiency, of course, depends not only on current management quality, but also on past management practices and decisions, as well as on earlier investment decisions. At the same time, the utility's social and economic environment plays an important role in the degree of efficiency it can attain, because local prices and regulation (including environmental and labor regulations), among other factors, affect efficiency levels. We will use two indicators to measure operational efficiency: nonrevenue water and staff productivity.

Our second key performance measure is *financial sustainability*. A utility that fails to cover at least its operation and maintenance costs from operating revenues is in a precarious position often leading to an inability to maintain infrastructure and to consequent deterioration in service quality. Even when operating revenues are sufficient to cover operating costs, however, a utility may still experience cash flow problems if customers do not pay their bills or pay them late. Therefore, the two indicators used here to determine financial sustainability are operating cost coverage ratio and collection period (the time it takes the utility to collect from its customers).

The utilities' *customer responsiveness* can be measured in many different ways. The indicator we use is affordability of service, as measured by how much of a household's income goes to water supply and sewerage services. Affordability also provides insight into the long-term sustainability of a utility; if its services are not affordable for its current population of consumers, the system will not be able to expand rapidly to serve larger, and often poorer, populations.

Water Coverage

Achieving the Millennium Development Goals for water supply and sanitation has been a major driver in the sector in the past decade. Between 2000 and 2007, median water supply coverage expanded from 81 percent in 2000 to 91 percent in 2008, despite rapidly increasing urban populations (table 3.1). (In this analysis, we will report median values, because using average values without considering the size of the utility will result in distortions as the performance of a small utility will count as much as that of a very large utility.) IBNET data have fluctuated over this period as increasing numbers of new utilities have entered the database. Usually, these newer utilities serve smaller and poorer populations than do the utilities that have been participating longer in the IBNET database. It is interesting to note the decline in standard deviation that assumes that the differences between utilities are declining over time.

Expansion of the IBNET database tends to have an adverse impact on performance, mostly because the larger the database, the greater the number of smaller utilities included. Smaller utilities tend to operate in smaller towns and, thus, to benefit less from economies of scale than do larger utilities.

Water-supply coverage, however, varies with income level. Utilities in low-income countries show lower water-supply coverage rates than do utilities in middle-income countries. In 2008, the median water coverage for households in low-income countries was 73 percent, compared to 91 percent in middle-income countries and 100 percent in high-income countries. Most of the increase in coverage has taken place in low-income countries, where median coverage increased by 14 percentage points, from 59 percent in 2000 to 73 percent in 2008, with much of this increase occurring in utilities in Africa.

Table 3.1   Median Coverage of Water-Supply Services

	2000	2001	2002	2003	2004	2005	2006	2007	2008
Water coverage (%)	81	82	98	89	90	90	91	91	91
Standard deviation (%)	25	25	60	24	23	23	23	22	22
Number of utilities reporting	637	700	803	1,086	1,242	1,223	1,432	1,296	989

Source: IBNET database.
Note: The data collection cycle for 2008 is not yet complete.

Wastewater Coverage

Median wastewater coverage increased from 54 percent in 2000 to 76 percent in 2008. As can be seen in table 3.2, the number of utilities providing wastewater services has increased rapidly. Nevertheless, wastewater coverage lags water-supply coverage. IBNET participation by wastewater service providers is also lower than participation by water-supply services.

Levels of wastewater coverage vary with the level of economic development. Utilities in low-income countries show lower rates of wastewater coverage than do utilities in middle-income countries. In 2008, average water coverage for households in low-income countries was 32 percent (in 2007), compared to 77 percent in middle-income countries and more than 95 percent in high-income countries. Wastewater coverage has increased most in middle-income countries, especially in Eastern Europe, where countries joining the European Union (EU) seek to comply with EU environmental standards.

As measured here, *wastewater coverage* refers to the collection of wastewater, not to the actual treatment or disposal of the wastewater collected. Nevertheless, levels of primary and secondary wastewater treatment increased between 2000 and 2008. In 2000, about 53 percent of utilities providing wastewater collection services also reported undertaking some level of primary treatment, but in 2008, 66 percent did so. Levels of secondary treatment have also increased, albeit less rapidly, with 28 percent of utilities in 2000 reporting some treatment of collected wastewater as compared to 31 percent in 2008.

Nonrevenue Water

Nonrevenue water (NRW) is calculated as the difference between water produced and water billed per kilometer of water network per day. This measure captures both physical and commercial losses. The latter result from inefficiencies in billing, illegal connections, and theft. High NRW levels indicate poor management, in the form of either poor commercial practices or poor infrastructure maintenance.

We will use several measures of NRW. The percentage of NRW as a share of water produced is a commonly used and easily understood indicator (table 3.3), but because it is very sensitive to changes in either of the two variables, we have found it to be unreliable for benchmarking NRW levels between utilities or even over time. This problem can be eliminated by measuring NRW not as a share, but in terms of absolute losses per kilometer of network or connection per day, as recommended by the International Water Association (IWA). Despite its shortcomings, the use of percentage figures to compare levels of NRW nevertheless remains common.

Table 3.2   Median Coverage of Wastewater Services

	2000	2001	2002	2003	2004	2005	2006	2007	2008
Wastewater coverage (%)	54	55	69	70	73	71	74	78	76
Standard deviation (%)	32	31	32	32	31	29	29	30	30
Number of utilities reporting	446	478	563	781	853	864	941	861	661

Source: IBNET database.
Note: The data collection cycle for 2008 is not yet complete.

Table 3.3 Nonrevenue Water (Percentage of Water Production)—Median Values

	2000	2001	2002	2003	2004	2005	2006	2007	2008
Nonrevenue water (%)	32	32	30	30	31	30	33	29	31
Standard deviation (%)	20	21	21	20	21	22	26	22	21
Number of utilities reporting	592	663	780	1,035	1,203	1,185	1,269	1,264	900

Source: IBNET database.
Note: The data collection cycle for 2008 is not yet complete.

The median nonrevenue water (as measured by the volume lost in percentage of water produced) has shown little progress between 2000 and 2008. Yet as can be seen in table 3.4, other measures of NRW show a different development pattern.

The median nonrevenue water (as measured by the volume lost in cubic meter per kilometer per day) has decreased from 27 in 2000 to 21 in 2008. But this indicator shows wide variations by year and between number of utilities (as shown in table 3.4). Progress has been made especially since 2004. Interestingly, the decline in NRW was accompanied by a decline in the standard deviation, assuming that the gap between utilities is also decreasing.

The data do not suggest a strong correlation between levels of NRW and economic development (see figure 3.1). On average, utilities in middle-income countries do not show any better management of NRW than do utilities in low-income countries. The median NRW in low-income countries was about 18 cubic meters per kilometer per day and about 22 in middle-income countries. In high-income countries (based on a relatively small group of observations), about 8 cubic meters per kilometer per day was lost in 2008. Many interlaced factors help explain NRW, including infrastructure age, network density, system pressure, and management quality.

Although NRW is lower in low-income countries, generally the median hours of supply is also significantly lower than in middle-income countries. In middle-income countries, the median utility offered 24 hours of water supply per day in 2008, compared to 16 hours per day in low-income countries. If 24 hours is considered the supply standard, only 16 percent of utilities in low-income countries complied with that standard in 2008, compared to 86 percent of utilities in middle-income countries.

Figure 3.2 shows the NRW in cubic meters per kilometer per day by size of utility. NRW tends to be lower in small utilities than in large utilities. One possible reason for this may be that smaller utilities are often relatively younger in age than larger utilities. Yet the biggest reason for the difference is likely to be that most larger utilities (those providing water-supply services to more than 500,000 people) generally serve more than one town and, hence, compose more than one water supply (and sewerage) system.

Staff Productivity

Fewer than half of the utilities in the IBNET sample provide information on staff productivity, as measured by the number of staff members per 1,000 connections. Those that have show improvement from 6.50 employees per 1,000 connections in 2000 to 3.26 in 2008 (see table 3.5). Yet, staff productivity varies widely from

Table 3.4   Nonrevenue Water (m³/km/day)—Median Values

	2000	2001	2002	2003	2004	2005	2006	2007	2008
Nonrevenue water (m³/km/day)	27	26	29	29	27	24	30	21	21
Standard deviation (m³/km/day)	73	80	84	62	55	51	56	50	50
Number of utilities reporting	605	635	720	962	1,060	1,059	1,096	1,204	869

Source: IBNET database.
Note: The data collection cycle for 2008 is not yet complete. m³/km/day = cubic meters per kilometer per day.

Figure 3.1   Nonrevenue Water (m³/km/day) by Income Level—Median Values

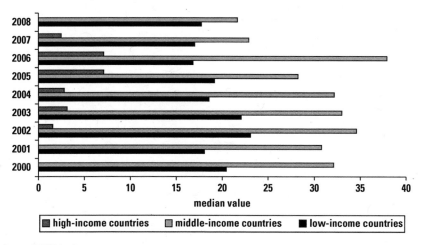

Source: IBNET database.
Note: The data collection cycle for 2008 is not yet complete. m³/km/day = cubic meters per kilometer per day.

about 20.0 employees per 1,000 connections in low-income countries to slightly above 3.0 in middle-income countries and about 0.80 in high-income countries. This variance in staff productivity is linked in part to differences in connection practices. In many places, water connections are often shared and, hence, serve multiple households. In Latin America, where most households have individual water connections, staff productivity is less than 3.0 per 1,000 connections, compared to 0.8 in high-income countries.

In Eastern Europe and Central Asia, many apartment buildings are still fitted with a single connection; in such an environment, staff productivity per 1,000 connections is likely to be very low, and in 2008, median staff productivity was about 12 employees per 1,000 connections. In Africa, staff productivity is low, partly because, as household surveys increasingly show, many households are not connected to the piped network but access the network by using (and often paying) for piped water supplied by neighbors. (Data from Demographic Health Surveys and Multiple Indicator Cluster Surveys underlie most of the data collected by the United Nations Children's Fund–World Health Organization Joint Monitoring Program, which measures progress toward the achievement of the Millennium Development Goals for water supply and sanitation.) Consequently,

Figure 3.2 Nonrevenue Water (m³/km/day) by Band Size of Utility (Measured by Number of People Served with Water Supply)—Median Values

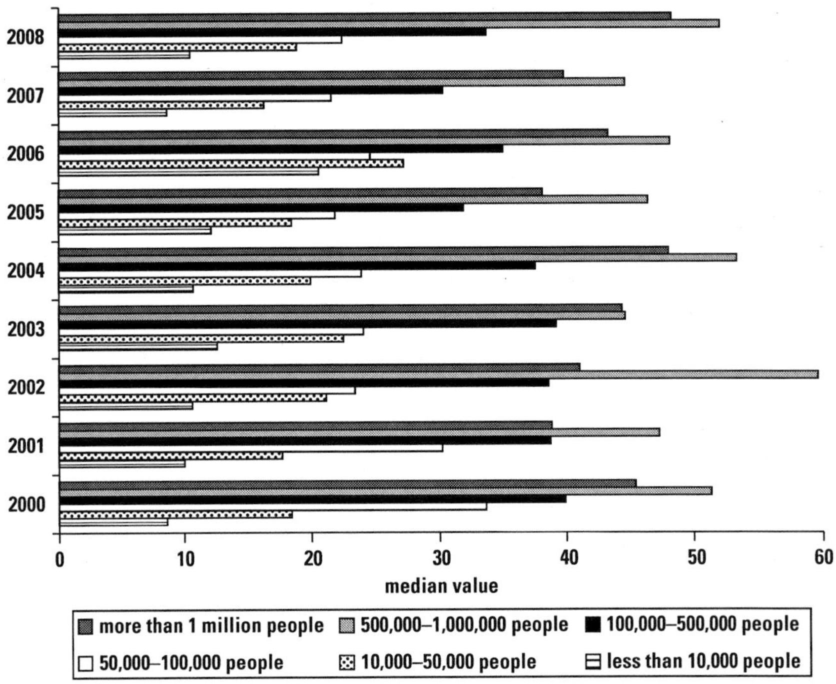

Source: IBNET database.
Note: The data collection cycle for 2008 is not yet complete. m³/km/day = cubic meters per kilometer per day.

Table 3.5 Median Staff Productivity

	2000	2001	2002	2003	2004	2005	2006	2007	2008
Staff productivity (employees per 1,000 people served)	1.39	1.35	1.47	1.16	1.06	1.06	0.95	1.49	0.97
Standard deviation (employees per 1,000 people served)	2.35	2.38	2.04	1.73	1.66	1.45	1.37	1.71	1.30
Number of utilities reporting	454	495	437	718	792	961	891	479	689

Source: IBNET database.
Note: The data collection cycle for 2008 is not yet complete.

median staff productivity was about 8 employees per 1,000 connections in 2008. But adjusting for the shared-connection effect, median staff productivity per 1,000 people served is much lower. Staff productivity stands at about 0.6 in Africa, 1.9 in Eastern Europe and Central Asia, and 0.8 in Latin America, compared to 0.2 in high-income countries.

Part of the increase in staff productivity may be attributable to outsourcing staff functions. In such cases, increased staff productivity does not necessarily translate into lower staff costs. In this respect, we see very divergent trends between regions. In Latin America and East Asia, labor costs decreased between

2000 and 2008, whereas the opposite happened in Africa and in Eastern Europe and Central Asia. Few high-income countries in the sample provide disaggregated details on their operating costs, leading to the conclusion that median labor cost as a percentage of total operating costs is about 30 percent in developed countries, compared to about 37 percent in developing countries. Labor share has not changed much globally, while staff productivity has increased, with the likely result that wages per employee have increased, providing incentives globally for improved performance by utility staff.

Labor costs as a percentage of total operating costs show very different trends in utilities providing water services only as compared to those providing both water and sewerage services. Utilities providing only water-supply services saw their share of labor costs in total operating costs decline from 45 percent in 2000 to 34 percent in 2008. At the same time, utilities providing both services saw their share of labor costs in total operating costs increase. The latter pattern is consistent with the increase in sewerage coverage discussed in the section "Wastewater Coverage" above. In 2008, utilities providing both water-supply and sewerage services had a median labor share of 38 percent, suggesting that utilities providing both services benefit from economies of scope.

Operating Cost Coverage Ratio

A utility's operating cost coverage ratio measures the extent to which revenues cover basic operation and maintenance costs. The median operating cost coverage ratio declined from 1.11 in 2000 to 1.05 in 2008, with most of that decrease taking place after 2003 (see table 3.6). Despite the triple crises in fuel, food, and financial markets, the impact on utilities has been negligible so far. Trends in the operating cost coverage ratio indicate, however, that even in the best of times the median utility barely covers its operation and maintenance costs, leaving it without the capacity to replace worn-out assets let alone expand services to larger groups of consumers.

The proportion of utilities unable to cover their basic operation and maintenance costs has increased from 35 percent in 2000 to 43 percent in 2008, with most of that increase occurring since the fuel crisis hit the sector (see figure 3.3). The effect is especially noticeable in low-income countries, where on average the percentage of utilities unable to cover even operation and maintenance costs increased most rapidly. Middle-income countries seem to be less affected, partially because many of these countries' economies continued to grow rapidly after 2004.

The operating cost coverage indicator measures both operating revenues and operating costs, but looks only at the basic indicators without providing details on the underlying trends in revenues and costs. The two elements that affect the operating cost coverage ratio are operation and maintenance costs and operating revenues. Each of these is affected by underlying trends in tariffs,

Table 3.6   Operating Cost Coverage Ratio—Median Values

	2000	2001	2002	2003	2004	2005	2006	2007	2008
Operating cost coverage ratio	1.11	1.13	1.10	1.11	1.08	1.07	1.07	1.08	1.05
Standard deviation	0.55	0.56	0.58	0.61	0.57	0.56	0.55	0.54	0.50
Number of utilities reporting	579	615	723	999	1,151	1,173	1,379	1,229	930

Source: IBNET database.
Note: The data collection cycle for 2008 is not yet complete.

Figure 3.3 Operating Cost Coverage Ratio—Median Values

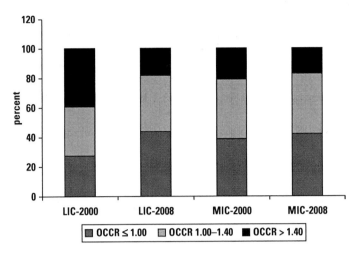

Source: IBNET database.
Note: The data collection cycle for 2008 is not yet complete. LIC = low-income countries. MIC = middle-income countries. OCCR = operating cost coverage ratio.

water consumption, and costs of inputs and by the efficiency with which they are applied.

Operation and Maintenance Costs. Median operation and maintenance (O&M) costs per cubic meter sold (expressed in U.S. dollars) have increased rapidly since 2000 and especially so since 2004 (see table 3.7). The large standard deviations suggest wide divergence between utilities in the cost of water and wastewater produced. The increased divergence between utilities is partially linked to exchange rate fluctuations, as some local currencies appreciated against the U.S. dollar in 2006 and 2007.

Operation and maintenance costs per cubic meter of water sold show wide variance between income levels. In low-income countries, operation and maintenance costs per cubic meter of water sold increased to US$0.23 in 2008, compared to US$0.68 in middle-income countries. The large variation in the levels of wastewater collection and treatment in low- and middle-income countries accounts for part of this difference. Other factors play a role as well, including general price levels in the countries. Moreover, public expenditure reviews of the water sector (for example, Tanzania) show that governments commonly fund part of the operation and maintenance costs by paying certain of them outright, thereby artificially depressing the cost of service.

As can be seen in figure 3.4, the median O&M cost per cubic meter of water sold has increased, particularly after 2004 when the full impact of the fuel crisis was felt. Interestingly, the variation in O&M cost also increased up until 2008, suggesting that different utilities responded differently to the fuel crisis. (Currency fluctuations can have a large impact on the development of IBNET indicators. Water service is paid for in local currency. When large currency fluctuations occur, the effect in U.S. dollar terms can be huge, an effect that can be explained as much or more by currency fluctuations than by cost increases.) Some countries,

Table 3.7   Operation and Maintenance Costs per Cubic Meter of Water Sold—Median Values

	2000	2001	2002	2003	2004	2005	2006	2007	2008
O&M cost (US$)	0.31	0.27	0.24	0.28	0.33	0.40	0.47	0.59	0.66
Standard deviation (US$)	0.39	0.37	0.34	0.35	0.41	0.46	0.52	0.56	0.46
Number of utilities reporting	541	580	697	949	1,103	1,128	1,188	1,201	872

Source: IBNET database.
Note: The data collection cycle for 2008 is not yet complete.

Figure 3.4   Operation and Maintenance Costs per Cubic Meter of Water Sold—Median Values

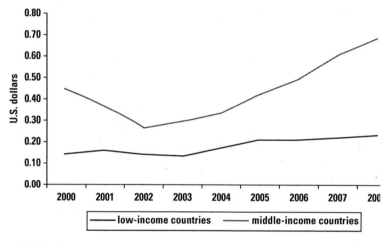

Source: IBNET database.
Note: The data collection cycle for 2008 is not yet complete.

especially emerging market economies, have seen their currencies rise against the U.S. dollar in very recent years; the increase in O&M costs thus dampened as fuel prices increased less rapidly in local currency terms, explaining the higher variance in increases in O&M costs per cubic meter of water sold among the different utilities.

The data also show that the proportion of energy costs in total operation and maintenance costs increased from 19 percent in 2000 to 23 percent in 2008, crowding out other costs. (The 2008 data are still being collected; as a result, the number of observations is relatively small. Only data with a sufficient number of observations will be reported.) Large differences appear in the figures for low-income and for middle-income countries.

In middle-income countries, the proportion of energy costs in total operation and maintenance cost has increased from 21 percent in 2003 to 23 percent in 2008. In low-income countries, however, energy costs as a proportion of total operation and maintenance costs decreased from 25 percent in 2003 to 20 percent in 2008. Utilities are clearly starting to consume less energy as energy prices increase. Two operational indicators account for this decline: per capita water production and duration of supply. Both indicators show declines in low-income countries. Median per capita water production has declined from 170 liters per capita per day (lcd) in 2003 to 96 lcd in 2008. Duration of supply shows a more gradual decline: the median for hours per day of supply was 16 in 2008.

Operating Revenues. Median revenues per cubic meter of water sold (as a proxy for tariffs) increased from US$0.37 in 2000 to US$0.71 in 2008 (see table 3.8). The increase in O&M costs is thus accompanied by an increase in revenues, suggesting that utilities have adjusted their prices to continue covering their O&M costs, albeit a little less rapidly. The increase in average revenues has been relatively limited in low-income countries (see figure 3.5).

In general, average revenues per cubic meter of water sold have increased across the board, independent of utility size. Average revenues per cubic meter of water sold, however, tend to be lowest in utilities serving fewer than 10,000 people and highest in utilities serving more than one million people.

Obviously, price increases will affect water consumption patterns. Water consumption in the past decade saw a sharp decline, especially in low-income countries, where median water consumption declined from 138 to 75 liters per capita per day between 2000 and 2008. In middle-income countries, water consumption

Table 3.8   Average Revenues per Cubic Meter of Water Sold—Median Values

	2000	2001	2002	2003	2004	2005	2006	2007	2008
Average revenues (US$)	0.37	0.34	0.28	0.32	0.37	0.43	0.50	0.63	0.71
Standard deviation (US$)	0.34	0.34	0.37	0.42	0.47	0.50	0.53	0.59	0.51
Number of utilities reporting	567	632	725	982	1,137	1,154	1,188	1,203	878

Source: IBNET database.
Note: The data collection cycle for 2008 is not yet complete.

Figure 3.5   Average Revenues per Cubic Meter of Water Sold—Median Values

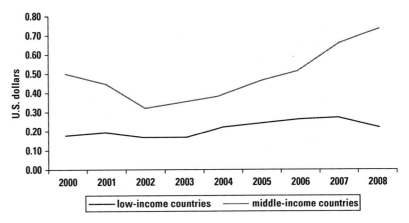

Source: IBNET database.
Note: The data collection cycle for 2008 is not yet complete.

shows a much more complex picture. Overall, water consumption remained more or less stable, but different regions show very different trends. In East Asia, water consumption increased; in Latin America, it more or less remained stable; and in Eastern Europe and Central Asia, water consumption declined. Higher incomes fueled by economic growth and differences in real tariff increases and metering policies help explain the variations in consumption. Between 2000 and 2008, East Asia benefited from fast economic growth and relatively small increases in water prices. During the same period, Latin America, Eastern Europe, and Central Asia were faced with much more rapidly increasing water prices. Although metering is widespread globally, it has increased rapidly since 2000 in Eastern Europe and Central Asia.

Collection Period

The median collection period decreased from 156 days in 2000 to 81 days in 2008. This is a rapid improvement and means that the median utility is achieving the 90-day benchmark. The indicator shows large variations between utilities, however. The IBNET database shows that long collection periods are particularly a problem in Asia. In general, collection periods tend to be longer in low-income countries than in middle-income countries, although the gap is declining.

Table 3.9 shows that the standard deviation is very high, indicating wide variation between utilities in the efficiency with which they collect billed revenues.

Affordability of Water and Sewerage Services

In many countries, affordability figures centrally in the discussion on water and sewerage services. Yet in the IBNET sample, the median affordability (measured as average revenues per capita as a percentage of gross national income [GNI] per capita) was 0.91 percent in 2008 (table 3.10). The actual numbers are likely to be smaller, because most utilities are serving urban populations (which tend to have higher average incomes than do rural populations), whereas the GNI per capita is a national average. The indicator varies considerably, however, with some households paying almost 6 percent for water and sewerage services. Between 2000 and 2008, median affordability improved from 1.09 percent to 0.91 percent.

In general, as might be expected, water and wastewater services are more costly than water-supply services only. Average affordability in 2008 was 1.16 percent for households using both water and wastewater services and 0.55 percent for those only using water-supply services. Affordability decreases with income levels. Utilities in low-income countries show higher rates of affordability than do

Table 3.9   Collection Period—Median Values

	2000	2001	2002	2003	2004	2005	2006	2007	2008
Collection period (number of days)	156	130	125	107	115	109	99	89	81
Standard deviation (number of days)	383	362	321	316	292	349	307	315	420
Number of utilities reporting	494	590	665	932	1,042	1,016	1,123	1,014	789

Source: IBNET database.
Note: The data collection cycle for 2008 is not yet complete.

Table 3.10   Affordability as Percentage of GNI—Median Values

	2000	2001	2002	2003	2004	2005	2006	2007	2008
Affordability (%)	1.09	0.98	0.93	1.00	0.96	0.96	0.94	1.12	0.91
Standard deviation (%)	1.67	4.76	3.98	4.43	1.18	1.14	1.02	0.98	1.00
Number of utilities reporting	613	676	757	1,026	1,184	1,183	1,378	1,255	937

Source: IBNET database.
Note: The data collection cycle for 2008 is not yet complete.

utilities in middle-income countries. In 2008, median affordability for households in low-income countries was 1.31 percent, compared to 0.90 percent in middle-income countries and 1.38 percent in high-income countries.

Figure 3.6 shows that affordability is highest in low-income countries, where consumers spent a larger part of their income on water-supply services. Service levels tend to increase with income levels, and households increasingly obtain access to wastewater services. Nevertheless, better service tends not to translate into less affordable service, because households in high-income countries usually spend less of their income on water supply and sewerage services than do households in low-income countries. Households in middle-income countries spend the least for these services. Obviously, access for the poor must be protected, because recent research shows that increasing infrastructure tariffs in combination with reform can increase income inequality (Milanovic and Ersado 2008).

Subsidies do not provide an easy solution to the problem of improving affordability, however. Most important, subsidies tend to be regressive, as a recent study on water and electricity subsidies amply demonstrated (Komives et al. 2005). The trend differs among countries. In almost all regions for which data are available, users depending on smaller quantities of water pay significantly more per cubic meter than do users of greater quantities. In part, this results from the high fixed costs characteristic of tariffs, which disproportionally affect those consuming smaller quantities of water. The high tariffs being paid in Africa are especially striking, especially because service levels tend to be relatively low there, thus confirming the conclusions of the Africa Infrastructure Country Diagnostic team that infrastructure in Africa is more expensive than in most other regions of the world.

Operating subsidies may also distort incentives for more efficient use of resources, and they tend to continue long after the crisis has subsided. In many water-supply and sewerage systems with far from universal coverage, operating subsidies tend to benefit consumers already connected to the piped-water system, who tend not to be the poor. The Africa Infrastructure Country Diagnostic studies (Foster 2008) showed that around 90 percent of the population with piped-water access belongs to the richest 60 percent of the population. Subsequently, in such an environment any subsidy to piped-water services is largely captured by better-off households.

Subsidies can be provided through social safety nets, direct subsidies to the water sector, or cross-subsidies. In the water sector, social safety nets are generally more effective than consumption or connection subsidies. Direct subsidies to the water sector require fiscal space in the sector budget, which may not necessarily be available. Cross-subsidies are easier to implement, and the direct fiscal repercussions are small, but the effectiveness of this instrument depends on the

Figure 3.6   Median Affordability as Percentage of GNI per Capita by Economic Development Status

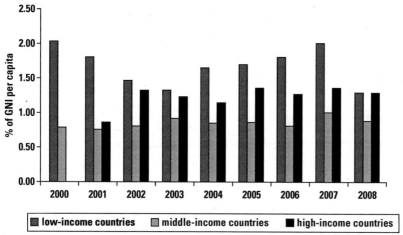

Source: IBNET database.
Note: The data collection cycle for 2008 is not yet complete.

Table 3.11   Level of Cross-Subsidies—Median Values

	2000	2001	2002	2003	2004	2005	2006	2007	2008
Affordability	2.06	2.17	2.00	1.75	1.80	1.73	1.60	1.62	1.35
Standard deviation	9.25	9.84	NA	9.20	9.65	8.98	13.81	9.20	11.75
Number of utilities reporting	351	320	346	389	507	487	503	540	254

Source: IBNET database.
Note: The data collection cycle for 2008 is not yet complete.

existing tariff structure. Maxing out cross-subsidies can be problematic: if the cost of water and wastewater becomes too expensive for nonresidential water users, they may opt out of the piped-water supply system, undermining the utilities' revenue base.

The IBNET database provides some details on the level of cross-subsidies in utilities, but it is incomplete, because many utilities do not provide this type of data, especially in Latin America. As can be seen in table 3.11, in 2008 the median utility charged nonresidential users up to 1.35 times more per cubic meter of water than it charged residential users. The large standard deviation, interestingly, shows that utilities display very different behaviors and that cross-subsidies vary widely between utilities. High levels of cross-subsidies tend to be more common in low-income countries than in middle-income countries.

The data also clearly show a direct relation between the level of cross-subsidies and the proportion of nonresidential water consumption in total water consumption. The higher the level of cross-subsidies, the higher the proportion of residential consumption and the lower the proportion of nonresidential water consumption in total water consumption. If the level of cross-subsidies is limited to less than one, nonresidential water consumption makes up 41 percent of total water consumption. If the level of cross-subsidies is between one and two, nonresidential water consumption drops to 27 percent; at a level of cross-subsidies of more than two it drops to 19 percent. High levels of nonresidential water consumption do not automatically translate into more revenues per cubic meter sold, however; an optimal level of relatively modest cross-subsidies (between one and two) optimizes the average revenues per cubic meter of water sold.

Conclusions

In the past five years, overall utility performance has improved, despite the impact of the triple crisis in fuel, food, and the financial markets. The analysis shows that progress has been made in reforming the water-utility sector: water rates have been increasing, and until the impacts of the fuel crisis were felt, the utilities' ability to cover at least their basic O&M costs had been improving. Other indicators also showed improvements, such as median staff productivity and the median collection periods, whereas median tariffs (in U.S. dollar terms) have increased.

According to the Global Economic Monitor database, fuel prices increased by 236 percent between 2003 and 2008, but utilities continued to cover their higher O&M costs with their operating revenues, thus guaranteeing the utility's short-term financial sustainability. (Short-term financial sustainability is defined as the capacity of a utility to cover its basic O&M costs with its operating revenues.) In 2003, the median operating cost coverage ratio for all utilities in the IBNET database stood at 1.11, while in 2008, the indicator stood at 1.05. So far, most utilities have been able to pass at least part, if not most, of the higher O&M costs through to consumers. This pass-through and the subsequently higher water tariffs have resulted in a decline in water consumption in many areas. This decline in median water consumption triggered by higher median water tariffs has been accompanied by a decrease in median water production and a decline in the median number of hours of supply per day.

The Water Utility Apgar Score

The term *Apgar score* originated in the system, developed by physician Virginia Apgar, for assessing the health of newborn infants quickly and summarily by assessing them on five simple criteria, giving them a score from zero to two for each, and classifying the totaled results according to a set scale. Our Apgar score for water-supply and sewage utilities does something similar, assessing the utilities' operational, financial, and social performance based on five or six indicators, depending on the type of service provided. Most other service analyses focus exclusively on financial and operational performance. But in many countries, utilities are judged not only on these criteria, but also on their effectiveness in delivering services to the population, including the poor. The simple set of criteria used in the utility Apgar score focuses on all three aspects of performance. The criteria are (i) water supply coverage; (ii) sewerage coverage; (iii) nonrevenue water; (iv) collection period; (vi) operating cost coverage ratio; and (vi) affordability of water and wastewater services. Each criterion is rated on a scale from zero to two, and the results are totaled. For utilities providing only water, the score is normalized (the maximum score for water utilities is 10; for water and wastewater utilities, 12). As with the original Apgar scores, utilities are then classified using a scale of overall viability: "critically low" utilities score 3.6 or less, "fairly low" utilities score between 3.6 and 7.2, and "normal" utilities score above 7.2.

The Apgar score, more particularly its set of indicators and benchmarks, is based on the characteristics of the IBNET database. Over time, the Apgar score will likely come to consist of different benchmarks and different indicators. As utilities develop, some indicators become less relevant, others more. In many developed countries, for example, service coverage is close to universal, making it less important as a measure of performance. Benchmarks may also change in value. For a sector showing improvement over time, benchmarks will likely require adjustment as well if the Apgar score is to remain relevant.

The average IBNET Apgar score was 7.06 in 2008, a fairly low overall score, but moving toward normal (table 3.12). The number of utilities in the green zone, that is, with Apgar scores classified as normal, has increased rapidly since 2000, and the number of utilities with performances classified as critically low has also decreased rapidly (figure 3.7).

Although an improvement in the average utility Apgar score was achieved between 2000 and 2008, the fuel crisis in 2004 resulted in a short-term setback: the IBNET Apgar score dropped, and the standard deviation increased. Since then, however, the score has increased steadily, especially after 2007. The financial crisis will also affect the utilities' performance, but the impact will likely be felt only after a delay, as with any revenue-side effect, given the grace period for bill paying allowed to households before services are cut.

Table 3.12   Classification of Water Utilities' Apgar Scores

	Indicator	Value	Average value of Apgar score for 2008
1.1	Water coverage	0 if ≤ 75% 1 if between 75 and 90% 2 if > 90%	1.21
2.1	Sewerage coverage	0 if ≤ 50% 1 if between 50 and 80% 2 if > 80%	1.17
6.2	Nonrevenue water	0 if ≥ 40 m^3/km/day 1 if ≥ 10 and < 40 m^3/km/day 2 if < 10 m^3/km/day	0.99
19.1	Affordability	0 if > 2.5% 1 if between 1.0% and 2.5% 2 if ≤1.0%	1.47
23.1	Collection period	0 if ≥ 180 days 1 if between 90 and 180 days 2 if < 90 days	1.33
24.1	Operating cost coverage ratio	0 if < 1 1 if between 1 and 1.40 2 if ≥ 1.40	0.74
	Overall Apgar score	Critically low ≤ 3.6 Fairly low 3.6–7.2 Normal > 7.2	7.06

Source: authors.
Note: The benchmarks are set on the basis of the database characteristics. The participation of an increasing number of utilities and change in the utilities' performance over time will likely affect the benchmarks; consequently, the benchmarks on which the utility Apgar scores are based will likely adjust over time.

Figure 3.7 Utility Apgar Score by Classification

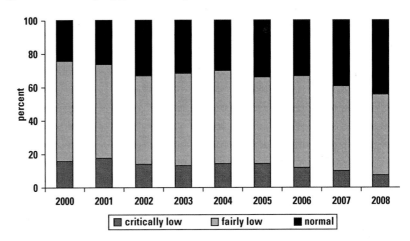

Source: IBNET database.
Note: The data collection cycle for 2008 is not yet complete.

While the Apgar score improved between 2000 and 2008 (with the exception of the years 2004 and 2005), the variance in performance between utilities has decreased, as measured by a decline in the standard deviation (see table 3.13). This convergence of performance is a global trend.

Despite the positive trend showing improvement in utility performance, large differences remain between utilities' Apgar scores and between countries. Utilities in low-income countries tend to have lower Apgar scores than do utilities in middle-income countries. Moreover, as table 3.14 shows, utilities in low-income countries as compared to those in middle-income countries tend to be more vulnerable to external shocks; the fuel crisis, for example, hit them harder.

As can be seen in figure 3.8, size also matters. Smaller utilities tend to have lower IBNET Apgar scores than do larger utilities. This only holds up to a point, however, because very large utilities are not necessarily the most efficient. Part of the performance difference arises when utilities serve more than one town. Utilities serving more than one town had an average Apgar score of 7.19 in 2008, compared to 6.54 for utilities serving only one town.

Interestingly, smaller utilities have made more progress in improving their performance than have larger utilities. Between 2000 and 2008, the smallest utilities, those serving fewer than 10,000 people, improved the most, while utilities serving more than one million people saw the least improvement.

One major reason why low-income countries score lower than do middle-income countries is because they tend to have lower levels of water and sewerage coverage. Yet very few utilities worldwide have been or are able to extend their coverage without public investments. Consequently, improvement in water and sewerage coverage tends to be dictated more by availability of public funding than by the utilities' ability to generate cash internally. This relationship becomes clearer when looking at a limited Apgar score measuring direct utility performance only, as expressed by the utility's capacity to control non-revenue water, collection periods, financial performance (as expressed in the

Table 3.13 Average Utility Apgar Score

	2000	2001	2002	2003	2004	2005	2006	2007	2008
IBNET Apgar	5.88	5.98	6.45	6.40	6.28	6.37	6.47	6.74	7.06
Standard deviation	2.37	2.39	2.31	2.25	2.27	2.29	2.19	2.17	2.09
Number of utilities reporting	437	494	571	783	886	838	830	931	684

Source: IBNET database.
Note: The data collection cycle for 2008 is not yet complete.

Table 3.14 Average Utility Apgar Scores by Level of Economic Development

	2000	2001	2002	2003	2004	2005	2006	2007	2008
Low-income countries	4.55	4.92	4.94	5.13	4.44	4.28	4.41	4.72	5.78
Middle-income countries	6.55	6.76	7.05	6.71	6.60	6.77	6.84	7.04	7.12

Source: IBNET database.
Note: The data collection cycle for 2008 is not yet complete.

**Figure 3.8 Apgar Score by Size of Utility
(Number of people served by water supply)**

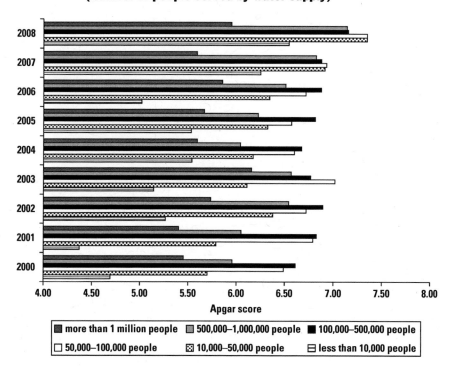

Source: IBNET database.
Note: The data collection cycle for 2008 is not yet complete.

operating cost coverage ratio), and affordability, normalized to obtain a score between 1 and 10.

Although the average utility management Apgar score improved between 2000 and 2007, as shown in table 3.15, the gap in management performance decreased. The fuel crisis in 2004 and the food crisis in 2007 constituted setbacks,

Table 3.15 Average Utility Management Apgar Scores by Level of Economic Development

	2000	2001	2002	2003	2004	2005	2006	2007	2008
Low-income countries	4.14	4.71	4.59	5.08	4.49	4.23	4.48	4.63	—
Middle-income countries	5.57	5.71	5.57	5.46	5.43	5.67	5.37	5.82	6.00
Apgar LIC as % of Apgar MIC	74	82	82	93	83	75	83	80	—

Source: IBNET database.
Note: The data collection cycle for 2008 is not yet complete. — = not available.

Box 3.1 Economies of Scale and Scope in Water Supply and Sewerage

Using panel data from the IBNET database, the study estimated measures of economies of scale and scope for four developing countries —Brazil, Moldova, Romania, and Vietnam—differing significantly in their levels of economic development, their piped-water and sewerage coverage, and the characteristics of their utilities. The study found evidence of economies of scale in three of the four countries (Moldova, Romania, and Vietnam), whereas the state water companies in Brazil showed constant returns of scale. Economies of scale were largest in Moldova (with the smallest utilities on average), and smallest in Romania. The study also found that returns to scale decrease with utility size. This result seems to hold up not only within countries, but also across countries. Finally, the study found evidence of economies of scope in the three countries in which utilities provide water and sewerage services (Brazil, Moldova, and Romania), showing that integrating provision of the two services has economic benefits.

Source: Nauges and van den Berg. 2008.

---

but the data seem to suggest that low-income countries are catching up and reducing the gap separating them from middle-income countries.

Size also matters when looking into the utility management Apgar scores. The smaller the utility, the higher the score. Whereas in 2008, the smallest utilities had a management Apgar score of 6.57, the largest utilities had scores of 4.93, compared to an average of 5.99. Looking at the impact of multi-systems, utilities serving more than one town had lower utility management Apgar scores than did those serving only one town. It is quite likely that these more complex multi-systems, which serve larger populations, require more management skills. Economies of scale thus play a role, as earlier studies have also shown (see box 3.1).

Appendix 1. From Benchmarking to Business Planning: The Case of Apa Canal Chisinau

The International Benchmarking Network for Water and Sanitation Utilities (IBNET) provides water and wastewater utilities with nearly unlimited possibilities for comparing their performance to that of other utilities around the world. But to what end? If a participating utility learns it has lower service coverage and cost-recovery rates than do many other utilities, should it conclude it is doing a poor job? Not necessarily, since local factors beyond the utility's control may well have been primarily responsible. Even where the comparison does point to a deficiency, what should the utility do? Benchmarking in itself does not put an organization on the path to improvement: taking action does.

But a good benchmarking system can provide a platform for organizational and service improvement. To participate in IBNET, a utility collects substantial amounts of basic current and historical data on its technical and economic performance. The data are loaded into the IBNET system, which automatically calculates a number of performance indicators. These indicators include many of the main building blocks for business planning, one of the best tools available to water and wastewater utilities for improving performance and achieving objectives. How can utilities use IBNET to begin a business planning process? What links IBNET's metric benchmarking to the reform-oriented, process benchmarking utilities can conduct in coordination with other water companies?

This appendix tries to answer those questions through a case study of Chisinau Apa Canal, the water and wastewater company in Chisinau, capital of Moldova. Chisinau Water was selected for the case study because it has both a long history of high-quality data with IBNET (see "IBNET Achievements" in chapter 1 of this book) and its own separate annual data covering its water and wastewater systems over a five-year period (2003–07).

Summary of Conceptual Framework for Business Planning

A business plan presents a detailed roadmap that can guide a utility from its current condition to a desired future state. Business plans often cover five-year periods and are tightly linked to the organization's annual budgeting process. The plan starts by assessing current conditions using existing information provided by the company's technical and economic departments. Next, the organization must define its objectives, answering the question "Where do we want to be in five years?" While all utilities have similar overall missions, development targets vary widely from company to company. The company's long-term mission is to provide all customers with reliable service at the lowest feasible cost while meeting quality and safety standards. But specific targets are particular to each institution: achieving 75 percent water-supply coverage may be an ambitious and worthwhile goal for one utility, while another company may already have 90 percent coverage and a goal of 95 percent. Financial targets too will vary: while one company may strive to use tariff revenues to cover only system operation and maintenance costs, another utility may be able to recover capital investment costs from end users, with or without loan financing. To maximize the value of

IBNET participation, given the varying needs of individual utilities, we must scale comparisons not to *all* utilities but to those at similar levels of development and with access to similar levels of financial resources.

Determining the best means of achieving objections is the greatest challenge of business planning. How will we finance the new wastewater treatment plant, given our customers' limited ability to pay? Can we cut operational costs to generate more financing for capital investment? What technical innovations might raise service levels? Although the specific challenges and solutions will vary from utility to utility, the analytical steps, shown in figure A1.1, are similar.

Within the business planning process, IBNET data and indicators are particularly helpful for demand forecasting, operations cost tracking, and calculation of revenue requirements. This case study presents and analyzes the relevant indicators for Chisinau Water in each of these areas. It also expands the analysis to the 10 largest water utilities in Moldova (of which Chisinau Water is the largest) and for the 10 largest utilities in four other countries with comparable levels of water- or wastewater-sector development: Ukraine, Romania, the Czech Republic, and Poland. Time-series cross-sectional analysis was used to determine the relationships between different variables and to test their statistical validity.

Demand Analysis

Demand should be the driver of any utility's program. Effective demand from current and future customers is the primary determinant of how much service to provide and when, where, and at what level. Utilities should carry out comprehensive demand analyses, detailed in table A1.1 below, including demographic analysis, water-use patterns, demand management, and wastewater demand.

Although IBNET does not provide all the data needed to complete a demand analysis, it does include many of the basic inputs. For the demographic analysis, for example, IBNET provides historical data on population and population served and calculates coverage levels as an indicator. For water-use patterns, IBNET provides data on water consumption, water sales, and water losses. In this case study, we show how IBNET data facilitate these analyses for Chisinau Water.

The population of Chisinau Water's service area has been relatively flat over the past 15 years. After increasing marginally in 2002–03, the population dropped to 700,000 before rising again to its current figure of 750,000. Over the same period, the population served by the water company has increased from 550,000 to more than 650,000. Using these two figures to calculate the percentage of population served, we see a slow but steady rise over the period to nearly 90 percent. Coverage equals (population served / total population) × 100.

Figure A1.2 shows the analysis graphically. This scatter chart is generated from the Excel spreadsheet the utility filled out when joining IBNET. Excel software produces such charts easily and quickly from its standard chart functions. Utility

Figure A1.1 Sequence of Analytical Steps

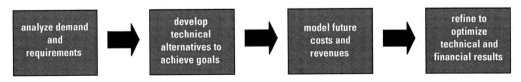

Source: Authors.

Table A1.1   Factors Included in Demand Analysis

Demand factor	Demand subfactor
Demographic analysis	• Service area by land-use patterns • Total population and population served by area, with cohort analysis • Housing structure, for example, single-family, multi-family, high-rise structures
Water-use patterns	• Historical water use by customer type: total in cubic meters and liters per capita per day, with trends • Forecasts by customer groups based on trends and land-use patterns • Weather impacts • Price-elasticity analysis
Demand management	• Conservation • Leak detection and remediation • Recycling • Unauthorized use
Wastewater demand	• Percentage of water sales entering the sewer system • Inflow and infiltration (I & I) • Influent to wastewater treatment plants (WWTPs) • Flow and strengths of plant influent • Discharge quality and methods

Source: Authors.

Figure A1.2   Trends in Population and Population Receiving Water Supply Services, Chisinau Water, 1994–2008

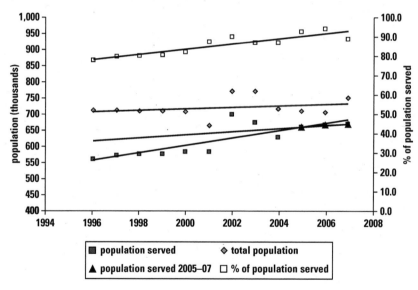

Source: IBNET.

analysts that do not use Excel can generate the same chart with pencil and paper by simply plotting the different points and then drawing the line best approximating the observed trend. Note that the average increase in population served slowed from 12,000 per year over the period 1996–2004 to only 5,000 per year since then. A slower growth rate is typical as the percentage of population served reaches 90 percent, as is the case in Chisinau.

Chisinau's coverage trends for wastewater are similar to those for its water system. Figure A1.3 below compares the population and population receiving wastewater services in the capital city. While about 11,600 people on average were added yearly to the wastewater population served, the total wastewater population increased by only 2,400 people per year. So the percentage of population served grew at an average annual rate of 1.4 percent.

Establishing the historical population trends in the service area placed Chisinau Water in a better position to project future population. Whether the projection will continue current trends or depart from them, dipping upward or downward, will depend on birth rates, death rates, and migration rates to and from the area. These in turn are influenced by macrolevel socioeconomic factors, such as actual and perceived local economic conditions, employment generation, the public-health system, and so on. The services of a demographer can be useful for refining these projections.

Before projecting water demand, we must first look at historical trends in water use. IBNET provides key data and indicators for this analysis. For Chisinau Water, we will examine only three: consumption of water in cubic meters by customer group, liters per capita per day (lcd) per customer group, and the relationship between water sales and water production.

Figure A1.4 shows total water sales, sales to residential customers, and sales to nonresidential customers. The latter two lines have different slopes over the period 2003–07: residential sales have been increasing by 1.75 million cubic meters per year, while nonresidential sales are actually decreasing by about 150,000 cubic meters annually. The reasons for lower nonresidential consumption could reflect closure of some industries, better demand management in the nonresidential sector, and or increasing access by industries to other sources of water. The trends in lcd are similar to those for total consumption per year for the different customer groups. Average daily water sales for all users is 206 lcd.

Figure A1.3 Trends in Population and Population Receiving Wastewater Services, Chisinau Water, 1994–2008

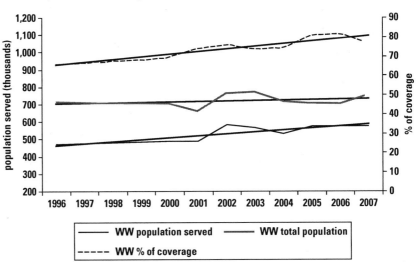

Source: IBNET.
Note: WW = wastewater.

Figure A1.4   Sales by Customer Group, Chisinau Water, 2003–07

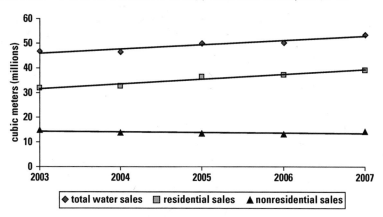

Source: IBNET.

As the figure demonstrates, Chisinau Water experienced declining nonindustrial sales, but overall sales still increased with time. This reflects the utility's success in rapidly expanding its percentage of population served. The rate of growth will slow as the percentage of population served approaches 100 percent.

It is useful to measure the relationship between population served and total water sales. One projection might be that cities with heavy industrial water use will have greater water sales per population served. Using IBNET data, we tested this hypothesis for five cities, selecting the largest utility (based on customers served) in each of five countries: Moldova, Romania, Ukraine, the Czech Republic, and Poland. After calculating for each utility the average population served and average total annual sales over the period 2003–07, we carried out a regression analysis of the relationship of population served to water sales. The results are shown in figure A1.5. If all of the cities were heavily residential and consumed about 150 liters per capita per day, we would expect this graph to show a fairly tight linear relationship related solely to population served; however, the relationship is nonlinear, with an upward slope to the curve increasing at the exponential rate of × 2. Thus, the larger the population served, the greater the water sales, following an exponential pattern. The reasons for this include, among others, the percentage residential versus industrial users, population density in large cities, household income, and availability of the water supply. The relative contributions of these factors were not evaluated.

The relationship is strong, with an R^2 of 0.955; that is, 95.5 percent of the change in sales is explained by change in the population served. The strength of the relationship may lend this graph some "predictive" powers: Chisinau Water could get a rough idea of future water sales by locating its future population served along the x-axis and using the regression line to locate the corresponding amount of water sold on the y-axis. This would also be true for utilities in other countries working under similar conditions and with comparable levels of population served. For identification of the causal factors for the exponential relationship, additional analysis would be required.

Another input into demand analysis calculable from IBNET data is nonrevenue water, that is, water lost to leakage, evaporation, unauthorized consumption,

Figure A1.5   Total Water Sales as a Function of Population Served for Five Utilities

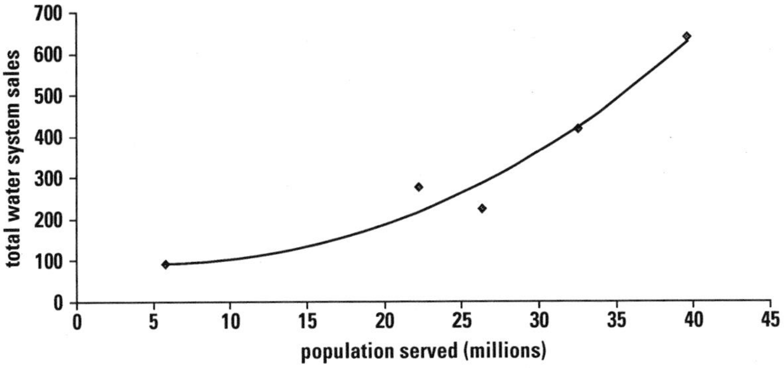

Source: IBNET.
Notes: Data points from left to right each represent the largest utility in Moldova, Romania, Ukraine, the Czech Republic, and Poland; values are calculated as averages for the period 2003–07 (for the Czech Republic, 2000–05).

and faulty metering. For a close approximation of nonrevenue water in Chisinau, we can examine the difference between water production and water sales (see figure A1.6). Water losses dropped from 36.4 million cubic meters (43.8 percent of production) in 2003 to 34.3 million cubic meters (39.0 percent of production) in 2007. Although decreasing, the level of water losses remains large and raises a red flag for the utility, highlighting an area that requires urgent action.

Figure A1.7, below, depicts the relationship of water production to sales for the largest utilities in Moldova, Romania, Ukraine, the Czech Republic, and Poland. At a given level of production (x-axis), predicted water sales can be derived by moving up to the regression line and then left to the y-axis (along the dashed line). The example indicates that for a production of about 44.5 million cubic meters, water sales would be about 30.5 million cubic meters, reflecting a water loss of about 14.0 million cubic meters, or 31.5 percent. This method is appropriate for evaluating any utility's water losses relative to the losses of its peers. The larger the sample size, the greater the reliability of the conclusions.

Operating Cost Analysis

When undertaking business planning, knowing the relationship between price and the variable cost of production, often referred to as the variable margin, is essential. The variable margin is the contribution of each additional unit of sales to profit (margin). Variable cost is derived using a simple scatter chart between operation costs (in U.S. dollars) and water sales volume. (Normally, the calculation involves production, not sales volume, but the real operating cost must also include the effects of lost water.)

The demand analysis discussed in the previous section allows development of future demand projections over the planning period. To meet future demand, the utility must optimize the operation and management of its existing water and wastewater system as well as make additional capital improvements as required. Both of these actions have cost and revenue implications. IBNET can provide the

Figure A1.6 Comparison of Water Production and Sales, Chisinau Water

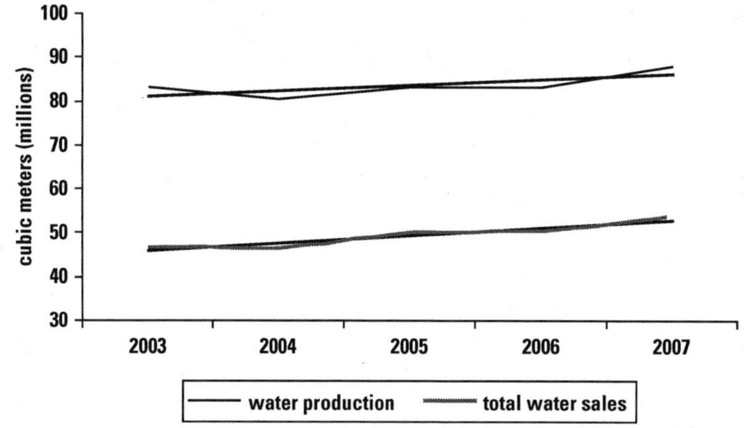

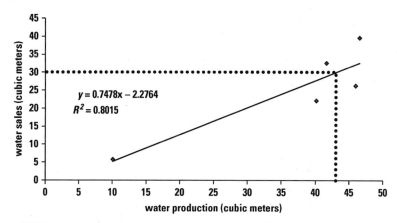

Source: IBNET.

Figure A1.7 Relationship of Total Water Sales to Water Production for Utilities in Five Capital Cities

$$y = 0.7478x - 2.2764$$
$$R^2 = 0.8015$$

Source: IBNET.
Note: Countries from left to right: Moldova, Romania, Ukraine, the Czech Republic, and Poland.

background for this analysis by representing the trends in operating costs and the tariff revenues needed to pay those costs. IBNET does not currently provide data on capital costs or debt financing.

Chisinau Water's total operating costs for the 2003–07 period are U-shaped (see figure A1.8). Operating costs fell in 2004, but they have been rising at ever-increasing annual rates since then. The sharp increase in 2007 may reflect rising energy costs.

IBNET can also be used to represent trends in operating costs per cubic meter produced or sold. Costs per cubic meter produced are significantly lower, since much more water is produced than is sold.

The regression line in figure A1.9 indicates operating costs are nearly flat over the given range in sales volume. (The slight decline must be attributed to the

Figure A1.8 Water System Operating Costs, Chisinau Water

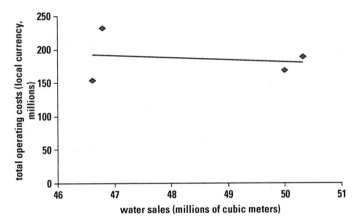

Source: IBNET.

Figure A1.9 Operating Cost as a Function of Amount of Water Sold

Source: IBNET.

small size of the sample.) A flat level of cost as the volume increases would indicate totally fixed costs. Water system operating costs are often 80 percent fixed, a realistic estimate for Chisinau. Chisinau's average operating cost per cubic meter is MDL 2.95. The average selling price for water is MDL 4.59; consequently, with a variable cost of MDL 0.59 (20 percent of 2.95), each additional unit sold would yield a variable or incremental margin of MDL 4.00 (4.59–0.59). This strongly suggests that, given an adequate water supply, distribution channels should be aggressively pursued to move plants up to near-full capacity utilization to maximize the margin they produce.

Other utilities participating in IBNET can create a similar scatter chart in Excel or on graph paper and calculate the extent to which increasing sales volume can improve the profitability of utility operations.

Calculation of Revenue Requirements

Revenue requirements vary with the utility's stage of development. As a general rule, all utilities should seek to cover operating costs through tariff collections. More mature utilities often try to collect enough tariff revenue to cover both operating and capital costs.

While IBNET does not provide data on capital costs or debt financing, it does provide tariff revenue data, which allows revenue analysis in relation to costs and volume of water sold. All monetary values in this section are in U.S. dollars, allowing comparisons with utilities in other countries.

As shown in figure A1.10, the unit price for water in Chisinau increased only moderately from 2003 to 2007, rising from US$0.30 per cubic meter to US$0.36, an average annual increase of 4.7 percent. Revenue increased more rapidly than price because of growth in volume (from US$14.00 in 2003 to US$19.10 in 2007, an increase of 8.1 percent per year).

Figure A1.11 explores the relationship between water volume and revenue in the five Eastern European countries. The second degree polynomial function yields an almost perfect fit for four of the five countries. One reason for higher revenues at higher volume might be the higher percentage of nonresidential customers in the larger cities or more developed countries. In many countries, tariffs for industrial and commercial (nonresidential) customers are structured to subsidize residential customers as a way to mitigate weak affordability among the general population. This is the case for two of the countries on the regression line. In a test of 2005 tariffs, Moldova and Ukraine have nonresidential tariffs that are, respectively, 5.23 and 2.76 times their residential tariff. The Czech Republic, Poland, and Romania have virtually uniform rates for residential and nonresidential customers. Eliminating the data from the two subsidizing countries would yield a virtual straight line for the regression curve. This illustrates that every country's tariff structure development is largely the result of unique

Figure A1.10   Water System Total Tariff Revenue and Average Price (US$), Chisinau

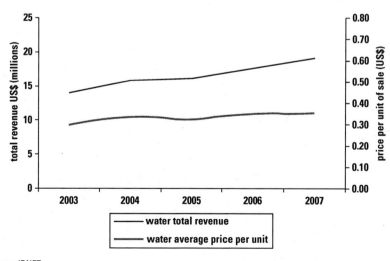

Source: IBNET.

Figure A1.11　Water Revenue Related to Water Sales Volume for Five Countries

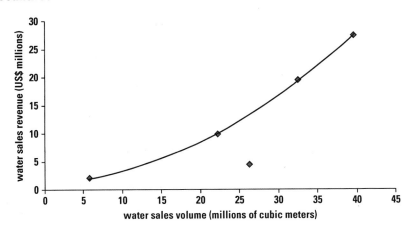

Source: IBNET.
Note: Countries from left to right: Moldova, Romania, Ukraine, the Czech Republic, and Poland.

policy situation. Some countries seek the simplicity of uniform tariffs, others the fairness of cost-of-service tariffs, and still others the financial assistance provided by cross-subsidies, especially during the early phases of development.

In considering water revenues in relation to water costs, we should recall, as noted above, that tariff revenues for start-up utilities should generally be equal to or larger than operating costs, not including debt service or depreciation. This yields a cost recovery ratio (CRR; tariff revenues/operating costs) of 1.0 or higher. For more mature utilities, the denominator should include both operating costs and capital costs (including debt servicing).

As shown in figure A1.12, Chisinau Water exceeded the target of 1.0 in three of the five years over the period 2003–07. Operating costs in both down years were significantly higher than in the up years, which led to the lower cost recovery ratio. Water volumes were steady. It is not critical that utilities achieve the ratio in every year. It is a target as well as a requirement for some loan covenants, but in start-up companies, achieving the CRR must be balanced with other goals and conditions, including affordable tariffs, the need to accommodate spikes in operating costs to meet urgent service requirements, and the level of subsidy available from government or donor organizations.

The average cost recovery ratio of the five regional utilities is given in figure A1.13, compared to a 1.0 performance. Three of the countries are above the targeted ratio of 1.0; the other two countries are only slightly below the target.

While an important indicator of financial performance, CRR is only one input into financial planning. The financial modeling in the course of business planning must evaluate all spending and revenue sources and work out a balance between them. To make ends meet, capital investments can be downsized, improvements can be pushed back in time, costs can be flattened through borrowing, and additional sources of funds can be mobilized. IBNET provides a platform for representing trends in costs and revenues on which utility managers can build a comprehensive financial model that takes these different factors into account.

Figure A1.12   Cost Recovery Ratio for the Water System, Chisinau

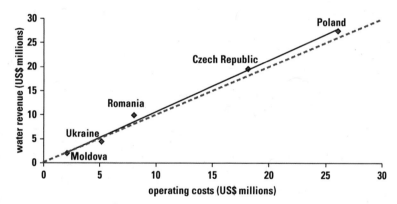

Source: IBNET.
Note: Dashed line = 1.0 CRR.

Figure A1.13   Water System Cost Recovery Ratio by Largest Utility in Five Countries

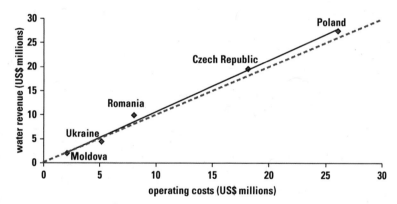

Source: IBNET.
Note: Dashed line = 1.0 CRR.

Summary of Key Points

- Benchmarking is a powerful management tool for all water and wastewater utilities. IBNET enables utilities to carry out metric benchmarking (comparing themselves to others and to their own performances over time) and provides a bridge between business planning and process benchmarking, using specific business processes that can be upgraded through cooperation with other similar utilities.

- IBNET data feed most directly into the business planning process in the areas of demand forecasting, operations cost tracking, and calculation of revenue requirements. The case study of Chisinau Water Company showed how trends in population served, water produced, water sold, operating costs, and tariff revenues can be represented graphically and used to provide a platform for comprehensive technical, financial, and economic analysis.

- Some of the statistical relationships discussed in this book, and the accompanying graphics, may be new to some IBNET program participants. All the statistical or numerical relationships presented, except the time-series, cross-sectional regressions, are basic and can be done using the standard Excel spreadsheet. Working out the numbers and taking them into account when making future plans is an essential part of strong, modern utility management. Moreover, quantitative analysis can have a major effect on the performance of a utility, both internally and in terms of the efficiency and quality of the services it provides. Finally, thorough quantitative analysis can be a powerful tool for convincing to government agencies and private investors of the need for additional financial resources to meet service delivery goals.

- Comparison of Chisinau Water with its counterparts in other Eastern European countries puts the utility in the context of others facing similar challenges. This demonstrates the importance of selecting the right peers: for a given utility, comparison with all utilities in the IBNET database would not be useful from the perspective of performance improvement. Rather, a utility should identify (perhaps using IBNET) a small number of utilities operating under similar conditions, with respect, for example, to population served, coverage, availability and quality of water, customers' ability to pay, available subsidies, and managerial and technical capacity. It is always advantageous to partner with a utility occupying a rung a little higher on the development ladder: we can learn a great deal from capable utilities that are already achieving what we want to achieve.

Appendix 2.   Country Data Tables

IBNET Indicator/Country: Albania

Latest year available	2007	2008	2009
Surface area (km^2)	28,748	28,748	28,748
GNI per capita, Atlas method (current US$)	2,200	2,400	3,950
Total population	3,132,458	3,143,291	3,200,200
Urban population (%)	46	47	47
Total urban population	1,443,437	1,468,546	1,450,000
MDGs			
Access to improved water sources, %, 2008 (WHO and UNICEF 2010)	97	97	97
Access to improved sanitation, %, 2008 (WHO and UNICEF 2010)	98	98	98
IBNET sourced data			
Number of utilities reporting in IBNET sample	55	55	56
Population served, water (thousands)	2,483	2,512	2,523
Size of the sample: total population living in service area, water supply (thousands)	3,109	3,112	3,160
Services coverage			
1.1 Water coverage (%)	77	77	79
2.1 Sewerage coverage (%)	46	45	61
Operational efficiency			
13.2 Electrical energy costs vs. operating costs (%) (share of energy cost as % of operational expenses)	25	—	26
6.1 Nonrevenue water (%)	69	71	70
6.2 Nonrevenue water (m^3/km/day)	88.70	91.20	85.20
12.3 Staff W/1,000 W population served (W/1,000 W population served)	—	2.00	2.00
15.1 Continuity of service (hours/day) (duration of water supply, hours)	9.95	10.27	10.09
Financial efficiency			
8.1 Water sold that is metered (%)	44	39	42
23.1 Collection period (days)	—	—	—
23.2 Collection ratio (%)	79	83	75
18.1 Average revenue W & WW (US$/m^3 water sold)	0.39	0.52	0.44
11.1 Operational cost W & WW (US$/m^3 water sold)	0.62	0.79	0.71
24.1 Operating cost coverage (ratio)	0.63	0.66	0.62
Production and consumption			
3.1 Water production (l/person/day)	303	301	300
4.1 Total water consumption (l/person/day)	105	94	97
4.7 Residential consumption (l/person/day)	76	74	78
Poverty and affordability			
19.1 Total revenues/service population GNI (% GNI per capita) (average revenues)	1	1	0.90
21.1 Ratio of industrial to residential tariff (level of cross-subsidy)	4.32	4.62	4.80

Source: All of the following Country Data Tables were compiled by the authors using the IBNET database.
Note: GNI = gross national income, km^2 = square kilometer, l = liter, m^3/km/day = cubic meter per kilometer per day, W = water WW = wastewater, — = not available.

IBNET Indicator/Country: Argentina

Latest year available	2004	2005	2006
Surface area (km^2)	2,780,400	2,780,400	—
GNI per capita, Atlas method (current US$)	3,580	4,300	6,000
Total population	—	38,731,603	39,105,347
Urban population (%)	91	91	92
Total urban population	—	35,400,685	35,820,498
MDGs			
Access to improved water sources, %, 2008 (WHO and UNICEF 2010)	97	97	97
Access to improved sanitation, %, 2008 (WHO and UNICEF 2010)	90	90	90
IBNET sourced data			
Number of utilities reporting in IBNET sample	8	8	17
Population served, water (thousands)	11,029	10,617	13,409
Size of the sample: total population living in service area, water supply (thousands)	12,547	12,145	15,784
Services coverage			
1.1 Water coverage (%)	88	87	85
2.1 Sewerage coverage (%)	68	66	63
Operational efficiency			
13.2 Electrical energy costs vs. operating costs (%) (share of energy cost as % of operational expenses)	8	10	12
6.1 Nonrevenue water (%)	33	31	31
6.2 Nonrevenue water (m^3/km/day)	72.00	69.10	59.90
12.3 Staff W/1,000 W population served (W/1,000 W population served)	0.30	0.30	0.50
15.1 Continuity of service (hours/day) (duration of water supply, hours)	24.00	24.00	24.00
Financial efficiency			
8.1 Water sold that is metered (%)	25	21	33
23.1 Collection period (days)	166	141	61
23.2 Collection ratio (%)	—	90	73
18.1 Average revenue W & WW (US$/m^3 water sold)	0.20	0.21	0.24
11.1 Operational cost W & WW (US$/m^3 water sold)	0.14	0.15	0.16
24.1 Operating cost coverage (ratio)	1.42	1.39	1.49
Production and consumption			
3.1 Water production (l/person/day)	417	369	398
4.1 Total water consumption (l/person/day)	362	371	340
4.7 Residential consumption (l/person/day)	183	166	62
Poverty and affordability			
19.1 Total revenues/service population GNI (% GNI per capita) (average revenues)	1	1	1
19.2 Annual bill for households consuming 6m^3 of water/month (US$/yr)	—	—	37.13
21.1 Ratio of industrial to residential tariff (level of cross-subsidy)	—	—	2.50

IBNET Indicator/Country: Armenia

Latest year available	2007	2008
Surface area (km²)	29,743	29,743
GNI per capita, Atlas method (current US$)	1,250	1,300
Total population	3,072,450	3,077,087
Urban population (%)	64	64
Total urban population	1,964,525	1,965,028
MDGs		
Access to improved water sources, %, 2008 (WHO and UNICEF 2010)	96	96
Access to improved sanitation, %, 2008 (WHO and UNICEF 2010)	90	90
IBNET sourced data		
Number of utilities reporting in IBNET sample	5	5
Population served, water (thousands)	1,724	1,752
Size of the sample: total population living in service area, water supply (thousands)	2,164	2,177
Services coverage		
1.1 Water coverage (%)	80	80
2.1 Sewerage coverage (%)	34	35
Operational efficiency		
13.2 Electrical energy costs vs. operating costs (%) (share of energy cost as % of operational expenses)	21	21
6.1 Nonrevenue water (%)	85	84
6.2 Nonrevenue water (m³/km/day)	108.80	94.70
12.3 Staff W/1,000 W population served (W/1,000 W population served)	1.60	1.60
15.1 Continuity of service (hours/day) (duration of water supply, hours)	12.00	13.20
Financial efficiency		
8.1 Water sold that is metered (%)	75	78
23.1 Collection period (days)	236	266
23.2 Collection ratio (%)	83	87
18.1 Average revenue W & WW (US$/m³ water sold)	0.41	0.47
11.1 Operational cost W & WW (US$/m³ water sold)	0.41	0.44
24.1 Operating cost coverage (ratio)	1.00	1.05
Production and consumption		
3.1 Water production (l/person/day)	606	598
4.1 Total water consumption (l/person/day)	146	151
4.7 Residential consumption (l/person/day)	92	94
Poverty and affordability		
19.1 Total revenues/service population GNI (% GNI per capita) (average revenues)	2	2
19.2 Annual bill for households consuming 6m³ of water/month (US$/yr)	28.13	32.10
21.1 Ratio of industrial to residential tariff (level of cross-subsidy)	1.46	1.46

IBNET Indicator/Country: Australia

Latest year available	2005	2006	2007
Surface area (km^2)	7,692,024	7,692,024	7,692,024
GNI per capita, Atlas method (current US$)	28,000	29,000	31,000
Total population	20,394,800	20,697,900	21,072,500
Urban population (%)	88	88	89
Total urban population	17,988,214	18,292,804	18,661,806
MDGs			
Access to improved water sources, %, 2008 (WHO and UNICEF 2010)	100	100	100
Access to improved sanitation, %, 2008 (WHO and UNICEF 2010)	100	100	100
IBNET sourced data			
Number of utilities reporting in IBNET sample	30	53	64
Population served, water (thousands)	18,950	20,374	21,295
Size of the sample: total population living in service area, water supply (thousands)	18,950	20,374	21,295
Services coverage			
1.1 Water coverage (%)	100	100	100
2.1 Sewerage coverage (%)	95	95	93
Operational efficiency			
13.2 Electrical energy costs vs. operating costs (%) (share of energy cost as % of operational expenses)	—	—	—
6.1 Nonrevenue water (%)	6	7	6
6.2 Nonrevenue water (m^3/km/day)	3.90	4.40	3.20
12.3 Staff W/1,000 W population served (W/1,000 W population served)	—	—	—
15.1 Continuity of service (hours/day) (duration of water supply, hours)	24.00	24.00	24.00
Financial efficiency			
8.1 Water sold that is metered (%)	100	100	100
23.1 Collection period (days)	—	—	—
23.2 Collection ratio (%)	—	—	—
18.1 Average revenue W & WW (US$/m^3 water sold)	1.96	2.15	2.70
11.1 Operational cost W & WW (US$/m^3 water sold)	0.84	1.11	1.54
24.1 Operating cost coverage (ratio)	2.60	2.04	1.84
Production and consumption			
3.1 Water production (l/person/day)	651	692	612
4.1 Total water consumption (l/person/day)	360	335	316
4.7 Residential consumption (l/person/day)	218	211	191
Poverty and affordability			
19.1 Total revenues/service population GNI (% GNI per capita) (average revenues)	—	—	—
19.2 Annual bill for households consuming 6m^3 of water/month (US$/yr)	—	—	—
21.1 Ratio of industrial to residential tariff (level of cross-subsidy)	1.00	1.00	1.00

IBNET Indicator/Country: Bangladesh

Latest year available	2007	2008	2009
Surface area (km²)	143,998	143,998	143,998
GNI per capita, Atlas method (current US$)	500	550	576
Total population	157,752,512	160,000,128	—
Urban population (%)	27	27	—
Total urban population	42,056,820	43,424,035	—
MDGs			
Access to improved water sources, %, 2008 (WHO and UNICEF 2010)	80	80	80
Access to improved sanitation, %, 2008 (WHO and UNICEF 2010)	53	53	53
IBNET sourced data			
Number of utilities reporting in IBNET sample	11	11	11
Population served, water (thousands)	11,203	12,195	13,135
Size of the sample: total population living in service area, water supply (thousands)	16,295	16,849	17,894
Services coverage			
1.1 Water coverage (%)	69	72	73
2.1 Sewerage coverage (%)	30	30	30
Operational efficiency			
13.2 Electrical energy costs vs. operating costs (%) (share of energy cost as % of operational expenses)	51	38	39
6.1 Nonrevenue water (%)	35	35	35
6.2 Nonrevenue water (m³/km/day)	156.30	166.10	181.50
12.3 Staff W/1,000 W population served (W/1,000 W population served)	0.30	0.30	0.30
15.1 Continuity of service (hours/day) (duration of water supply, hours)	8.68	9.09	9.12
Financial efficiency			
8.1 Water sold that is metered (%)	72	76	74
23.1 Collection period (days)	351	347	322
23.2 Collection ratio (%)	102	93	107
18.1 Average revenue W & WW (US$/m³ water sold)	0.13	0.13	0.14
11.1 Operational cost W & WW (US$/m³ water sold)	0.09	0.09	0.10
24.1 Operating cost coverage (ratio)	1.38	1.41	1.36
Production and consumption			
3.1 Water production (l/person/day)	674	733	797
4.1 Total water consumption (l/person/day)	107	108	108
4.7 Residential consumption (l/person/day)	96	97	97
Poverty and affordability			
19.1 Total revenues/service population GNI (% GNI per capita) (average revenues)	1	1	1
19.2 Annual bill for households consuming 6m³ of water/month (US$/yr)	7.40	7.42	7.39
21.1 Ratio of industrial to residential tariff (level of cross-subsidy)	3.79	3.66	3.89

IBNET Indicator/Country: Belarus

Latest year available	2006	2007	2008
Surface area (km²)	207,600	207,600	207,600
GNI per capita, Atlas method (current US$)	2,300	2,400	2,500
Total population	9,732,500	9,702,000	9,680,850
Urban population (%)	73	73	73
Total urban population	7,067,742	7,086,341	7,111,552
MDGs			
Access to improved water sources, %, 2008 (WHO and UNICEF 2010)	100	100	100
Access to improved sanitation, %, 2008 (WHO and UNICEF 2010)	93	93	93
IBNET sourced data			
Number of utilities reporting in IBNET sample	30	13	13
Population served, water (thousands)	3,335	1,165	1,185
Size of the sample: total population living in service area, water supply (thousands)	3,526	1,265	1,281
Services coverage			
1.1 Water coverage (%)	95	92	93
2.1 Sewerage coverage (%)	80	79	80
Operational efficiency			
13.2 Electrical energy costs vs. operating costs (%) (share of energy cost as % of operational expenses)	22	29	24
6.1 Nonrevenue water (%)	18	15	18
6.2 Nonrevenue water (m³/km/day)	22.00	16.50	16.80
12.3 Staff W/1,000 W population served (W/1,000 W population served)	1.50	1.40	1.40
15.1 Continuity of service (hours/day) (duration of water supply, hours)	23.88	24.00	24.00
Financial efficiency			
8.1 Water sold that is metered (%)	83	81	84
23.1 Collection period (days)	84	47	37
23.2 Collection ratio (%)	91	84	87
18.1 Average revenue W & WW (US$/m³ water sold)	0.57	0.75	0.92
11.1 Operational cost W & WW (US$/m³ water sold)	0.46	0.61	0.80
24.1 Operating cost coverage (ratio)	1.25	1.23	1.16
Production and consumption			
3.1 Water production (l/person/day)	385	400	410
4.1 Total water consumption (l/person/day)	259	241	208
4.7 Residential consumption (l/person/day)	197	183	156
Poverty and affordability			
19.1 Total revenues/service population GNI (% GNI per capita) (average revenues)	2	3	3
19.2 Annual bill for households consuming 6m³ of water/month (US$/yr)	16.66	25.53	34.27
21.1 Ratio of industrial to residential tariff (level of cross-subsidy)	14.10	10.34	8.95

IBNET Indicator/Country: Benin

Latest year available	2007	2008	2009
Surface area (km²)	112,622	112,622	112,622
GNI per capita, Atlas method (current US$)	660	700	750
Total population	8,128,208	8,267,626	8,328,208
Urban population (%)	40	40	40
Total urban population	3,251,000	3,307,050	3,331,000
MDGs			
Access to improved water sources, %, 2008 (WHO and UNICEF 2010)	75	75	75
Access to improved sanitation, %, 2008 (WHO and UNICEF 2010)	12	12	12
IBNET sourced data			
Number of utilities reporting in IBNET sample	1	1	1
Population served, water (thousands)	1,598	1,703	1,860
Size of the sample: total population living in service area, water supply (thousands)	3,070	3,170	3,270
Services coverage			
1.1 Water coverage (%)	58	52	54
2.1 Sewerage coverage (%)	3	3	3
Operational efficiency			
13.2 Electrical energy costs vs. operating costs (%) (share of energy cost as % of operational expenses)	21	15	19
6.1 Nonrevenue water (%)	28	24	28
6.2 Nonrevenue water (m³/km/day)	5.98	5.34	6.38
12.3 Staff W/1,000 W population served (W/1,000 W population served)	0.30	0.40	0.50
15.1 Continuity of service (hours/day) (duration of water supply, hours)	21.00	24.00	24.00
Financial efficiency			
8.1 Water sold that is metered (%)	100	100	100
23.1 Collection period (days)	190	219	199
23.2 Collection ratio (%)	100	93	91
18.1 Average revenue W & WW (US$/m³ water sold)	1.17	1.28	1.37
11.1 Operational cost W & WW (US$/m³ water sold)	0.74	0.78	0.70
24.1 Operating cost coverage (ratio)	1.58	1.64	1.97
Production and consumption			
3.1 Water production (l/person/day)	58	57	57
4.1 Total water consumption (l/person/day)	42	45	41
4.7 Residential consumption (l/person/day)	33	35	34
Poverty and affordability			
19.1 Total revenues/service population GNI (% GNI per capita) (average revenues)	3	4	4
19.2 Annual bill for households consuming 6m³ of water/month (US$/yr)	44.04	45.50	50.40
21.1 Ratio of industrial to residential tariff (level of cross-subsidy)	1.05	1.11	1.04

IBNET Indicator/Country: Bhutan

Latest year available	2002	2003	2004
Surface area (km²)	38,394	38,394	38,394
GNI per capita, Atlas method (current US$)	590	720	760
Total population	—	—	—
Urban population (%)	28	29	30
Total urban population	—	—	—
MDGs			
Access to improved water sources, %, 2008 (WHO and UNICEF 2010)	92	92	92
Access to improved sanitation, %, 2008 (WHO and UNICEF 2010)	65	65	65
IBNET sourced data			
Number of utilities reporting in IBNET sample	1	1	1
Population served, water (thousands)	40	42	43
Size of the sample: total population living in service area, water supply (thousands)	60	60	60
Services coverage			
1.1 Water coverage (%)	67	70	72
2.1 Sewerage coverage (%)	0	0	0
Operational efficiency			
13.2 Electrical energy costs vs. operating costs (%) (share of energy cost as % of operational expenses)	—	—	—
6.1 Nonrevenue water (%)	38	47	46
6.2 Nonrevenue water (m³/km/day)	47.50	68.80	68.00
12.3 Staff W/1,000 W population served (W/1,000 W population served)	1.40	1.30	1.30
15.1 Continuity of service (hours/day) (duration of water supply, hours)	13.00	13.00	13.00
Financial efficiency			
8.1 Water sold that is metered (%)	—	—	—
23.1 Collection period (days)	—	—	—
23.2 Collection ratio (%)	81	73	88
18.1 Average revenue W & WW (US$/m³ water sold)	0.04	0.06	0.06
11.1 Operational cost W & WW (US$/m³ water sold)	0.03	0.04	0.04
24.1 Operating cost coverage (ratio)	1.28	1.55	1.55
Production and consumption			
3.1 Water production (l/person/day)	240	241	250
4.1 Total water consumption (l/person/day)	156	150	151
4.7 Residential consumption (l/person/day)	105	101	102
Poverty and affordability			
19.1 Total revenues/service population GNI (% GNI per capita) (average revenues)	0	0	0
19.2 Annual bill for households consuming 6m³ of water/month (US$/yr)	—	—	—
21.1 Ratio of industrial to residential tariff (level of cross-subsidy)	0.95	0.83	0.82

IBNET Indicator/Country: Bolivia

Latest year available	2004	2005	2006
Surface area (km^2)	1,098,581	1,098,581	1,098,581
GNI per capita, Atlas method (current US$)	960	1,000	1,020
Total population	—	9,182,062	9,353,826
Urban population (%)	64	64	65
Total urban population	—	5,894,884	6,048,184
MDGs			
Access to improved water sources, %, 2008 (WHO and UNICEF 2010)	86	86	86
Access to improved sanitation, %, 2008 (WHO and UNICEF 2010)	25	25	25
IBNET sourced data			
Number of utilities reporting in IBNET sample	2	2	5
Population served, water (thousands)	2,321	2,388	2,155
Size of the sample: total population living in service area, water supply (thousands)	2,355	2,510	2,453
Services coverage			
1.1 Water coverage (%)	99	95	88
2.1 Sewerage coverage (%)	69	64	66
Operational efficiency			
13.2 Electrical energy costs vs. operating costs (%) (share of energy cost as % of operational expenses)	—	6	23
6.1 Nonrevenue water (%)	28	28	35
6.2 Nonrevenue water (m^3/km/day)	13.10	17.40	24.40
12.3 Staff W/1,000 W population served (W/1,000 W population served)	—	0.20	0.80
15.1 Continuity of service (hours/day) (duration of water supply, hours)	24.00	24.00	20.00
Financial efficiency			
8.1 Water sold that is metered (%)	—	100	92
23.1 Collection period (days)	—	117	72
23.2 Collection ratio (%)	—	91	723
18.1 Average revenue W & WW (US$/m^3 water sold)	0.67	0.45	0.40
11.1 Operational cost W & WW (US$/m^3 water sold)	0.58	0.44	0.26
24.1 Operating cost coverage (ratio)	1.31	1.02	1.56
Production and consumption			
3.1 Water production (l/person/day)	85	113	100
4.1 Total water consumption (l/person/day)	72	93	83
4.7 Residential consumption (l/person/day)	94	78	61
Poverty and affordability			
19.1 Total revenues/service population GNI (% GNI per capita) (average revenues)	—	2	1
19.2 Annual bill for households consuming 6m^3 of water/month (US$/yr)	—	—	25.88
21.1 Ratio of industrial to residential tariff (level of cross-subsidy)	—	—	3.26

IBNET Indicator/Country: Bosnia and Herzegovina

Latest year available	2005	2006	2007
Surface area (km^2)	51,209	51,209	51,209
GNI per capita, Atlas method (current US$)	2,100	2,200	2,400
Total population	3,781,274	3,781,488	3,778,410
Urban population (%)	46	46	47
Total urban population	1,728,042	1,750,073	1,770,563
MDGs			
Access to improved water sources, %, 2008 (WHO and UNICEF 2010)	99	99	99
Access to improved sanitation, %, 2008 (WHO and UNICEF 2010)	95	95	95
IBNET sourced data			
Number of utilities reporting in IBNET sample	21	22	22
Population served, water (thousands)	1,215	1,221	1,266
Size of the sample: total population living in service area, water supply (thousands)	1,337	1,315	1,364
Services coverage			
1.1 Water coverage (%)	91	93	93
2.1 Sewerage coverage (%)	57	57	56
Operational efficiency			
13.2 Electrical energy costs vs. operating costs (%) (share of energy cost as % of operational expenses)	13	13	119
6.1 Nonrevenue water (%)	61	62	61
6.2 Nonrevenue water (m^3/km/day)	74.30	61.30	58.30
12.3 Staff W/1,000 W population served (W/1,000 W population served)	1.40	1.30	1.30
15.1 Continuity of service (hours/day) (duration of water supply, hours)	23.24	23.33	24.00
Financial efficiency			
8.1 Water sold that is metered (%)	99	98	99
23.1 Collection period (days)	239	257	343
23.2 Collection ratio (%)	79	83	159
18.1 Average revenue W & WW (US$/m^3 water sold)	0.72	0.77	0.82
11.1 Operational cost W & WW (US$/m^3 water sold)	0.60	0.80	0.84
24.1 Operating cost coverage (ratio)	1.05	0.94	0.97
Production and consumption			
3.1 Water production (l/person/day)	204	188	189
4.1 Total water consumption (l/person/day)	178	159	161
4.7 Residential consumption (l/person/day)	134	118	119
Poverty and affordability			
19.1 Total revenues/service population GNI (% GNI per capita) (average revenues)	2	2	2
19.2 Annual bill for households consuming 6m^3 of water/month (US$/yr)	44.86	46.57	53.71
21.1 Ratio of industrial to residential tariff (level of cross-subsidy)	2.79	2.86	2.66

IBNET Indicator/Country: Brazil

Latest year available	2006	2007	2008
Surface area (km²)	8,514,877	8,514,877	8,514,877
GNI per capita, Atlas method (current US$)	3,350	3,800	4,700
Total population	188,158,438	190,119,995	191,971,506
Urban population (%)	85	85	86
Total urban population	159,294,934	161,830,140	164,289,215
MDGs			
Access to improved water sources, %, 2008 (WHO and UNICEF 2010)	97	97	97
Access to improved sanitation, %, 2008 (WHO and UNICEF 2010)	80	80	80
IBNET sourced data			
Number of utilities reporting in IBNET sample	592	605	661
Population served, water (thousands)	140,941	141,149	146,392
Size of the sample: total population living in service area, water supply (thousands)	178,069	176,968	182,107
Services coverage			
1.1 Water coverage (%)	79	80	81
2.1 Sewerage coverage (%)	41	42	43
Operational efficiency			
13.2 Electrical energy costs vs. operating costs (%) (share of energy cost as % of operational expenses)	22	29	25
6.1 Nonrevenue water (%)	41	40	39
6.2 Nonrevenue water (m³/km/day)	35.60	34.10	33.50
12.3 Staff W/1,000 W population served (W/1,000 W population served)	0.50	—	—
15.1 Continuity of service (hours/day) (duration of water supply, hours)	—	24.00	24.00
Financial efficiency			
8.1 Water sold that is metered (%)	77	94	76
23.1 Collection period (days)	147	115	112
23.2 Collection ratio (%)	95	92	99
18.1 Average revenue W & WW (US$/m³ water sold)	1.17	1.49	1.56
11.1 Operational cost W & WW (US$/m³ water sold)	0.82	1.38	1.04
24.1 Operating cost coverage (ratio)	1.43	1.08	1.49
Production and consumption			
3.1 Water production (l/person/day)	200	213	220
4.1 Total water consumption (l/person/day)	162	169	167
4.7 Residential consumption (l/person/day)	—	—	—
Poverty and affordability			
19.1 Total revenues/service population GNI (% GNI per capita) (average revenues)	1	1	1
19.2 Annual bill for households consuming 6m³ of water/month (US$/yr)	—	—	—
21.1 Ratio of industrial to residential tariff (level of cross-subsidy)	1.00	1.00	1.00

IBNET Indicator/Country: Bulgaria

Latest year available	2006	2007	2008
Surface area (km^2)	110,879	110,879	110,879
GNI per capita, Atlas method (current US$)	3,100	3,150	3,200
Total population	7,699,020	7,659,764	7,623,395
Urban population (%)	71	71	71
Total urban population	5,427,809	5,423,113	5,420,234
MDGs			
Access to improved water sources, %, 2008 (WHO and UNICEF 2010)	100	100	100
Access to improved sanitation, %, 2008 (WHO and UNICEF 2010)	100	100	100
IBNET sourced data			
Number of utilities reporting in IBNET sample	20	20	20
Population served, water (thousands)	5,246	5,398	5,389
Size of the sample: total population living in service area, water supply (thousands)	5,288	5,436	5,422
Services coverage			
1.1 Water coverage (%)	99	99	99
2.1 Sewerage coverage (%)	59	60	60
Operational efficiency			
13.2 Electrical energy costs vs. operating costs (%) (share of energy cost as % of operational expenses)	22	29	24
6.1 Nonrevenue water (%)	59	57	55
6.2 Nonrevenue water (m^3/km/day)	31.20	28.60	27.10
12.3 Staff W/1,000 W population served (W/1,000 W population served)	1.60	1.60	1.60
15.1 Continuity of service (hours/day) (duration of water supply, hours)	23.96	23.96	23.96
Financial efficiency			
8.1 Water sold that is metered (%)	98	98	99
23.1 Collection period (days)	145	136	106
23.2 Collection ratio (%)	129	123	120
18.1 Average revenue W & WW (US$/m^3 water sold)	0.68	0.78	1.00
11.1 Operational cost W & WW (US$/m^3 water sold)	0.50	0.58	0.77
24.1 Operating cost coverage (ratio)	1.38	1.35	1.32
Production and consumption			
3.1 Water production (l/person/day)	796	770	751
4.1 Total water consumption (l/person/day)	172	170	170
4.7 Residential consumption (l/person/day)	152	146	145
Poverty and affordability			
19.1 Total revenues/service population GNI (% GNI per capita) (average revenues)	1	1	2
19.2 Annual bill for households consuming 6m^3 of water/month (US$/yr)	45.04	56.24	72.39
21.1 Ratio of industrial to residential tariff (level of cross-subsidy)	0.94	0.92	0.89

IBNET Indicator/Country: Burkina Faso

Latest year available	2004	2005	2006
Surface area (km²)	274,222	274,222	274,222
GNI per capita, Atlas method (current US$)	350	370	400
Total population	—	13,747,182	14,224,581
Urban population (%)	18	18	19
Total urban population	—	2,515,734	2,662,842
MDGs			
Access to improved water sources, %, 2008 (WHO and UNICEF 2010)	76	76	76
Access to improved sanitation, %, 2008 (WHO and UNICEF 2010)	11	11	11
IBNET sourced data			
Number of utilities reporting in IBNET sample	1	1	1
Population served, water (thousands)	2,300	2,780	2,930
Size of the sample: total population living in service area, water supply (thousands)	2,640	3,047	3,135
Services coverage			
1.1 Water coverage (%)	87	91	93
2.1 Sewerage coverage (%)	0	0	0
Operational efficiency			
13.2 Electrical energy costs vs. operating costs (%) (share of energy cost as % of operational expenses)	13	11	13
6.1 Nonrevenue water (%)	22	23	24
6.2 Nonrevenue water (m³/km/day)	7.00	7.80	7.90
12.3 Staff W/1,000 W population served (W/1,000 W population served)	0.30	0.30	0.30
15.1 Continuity of service (hours/day) (duration of water supply, hours)	—	—	—
Financial efficiency			
8.1 Water sold that is metered (%)	100	100	100
23.1 Collection period (days)	—	—	—
23.2 Collection ratio (%)	105	95	105
18.1 Average revenue W & WW (US$/m³ water sold)	1.03	1.04	1.13
11.1 Operational cost W & WW (US$/m³ water sold)	1.23	1.30	1.27
24.1 Operating cost coverage (ratio)	0.83	0.80	0.89
Production and consumption			
3.1 Water production (l/person/day)	37	41	45
4.1 Total water consumption (l/person/day)	35	31	32
4.7 Residential consumption (l/person/day)	31	28	29
Poverty and affordability			
19.1 Total revenues/service population GNI (% GNI per capita) (average revenues)	4	3	3
19.2 Annual bill for households consuming 6m³ of water/month (US$/yr)	55.79	55.87	58.94
21.1 Ratio of industrial to residential tariff (level of cross-subsidy)	9.78	8.35	7.38

IBNET Indicator/Country: Burundi

Latest year available	2004	2005	2006
Surface area (km^2)	27,834	27,834	27,834
GNI per capita, Atlas method (current US$)	90	100	110
Total population	—	7,378,129	7,603,492
Urban population (%)	9	10	10
Total urban population	—	700,922	745,142
MDGs			
Access to improved water sources, %, 2008 (WHO and UNICEF 2010)	72	72	72
Access to improved sanitation, %, 2008 (WHO and UNICEF 2010)	46	46	46
IBNET sourced data			
Number of utilities reporting in IBNET sample	1	1	1
Population served, water (thousands)	650	700	750
Size of the sample: total population living in service area, water supply (thousands)	6,000	6,500	7,000
Services coverage			
1.1 Water coverage (%)	11	11	11
2.1 Sewerage coverage (%)	—	—	—
Operational efficiency			
13.2 Electrical energy costs vs. operating costs (%) (share of energy cost as % of operational expenses)	—	—	—
6.1 Nonrevenue water (%)	45	40	40
6.2 Nonrevenue water (m^3/km/day)	19.80	17.30	17.20
12.3 Staff W/1,000 W population served (W/1,000 W population served)	0.60	0.60	0.70
15.1 Continuity of service (hours/day) (duration of water supply, hours)	15.00	15.00	15.00
Financial efficiency			
8.1 Water sold that is metered (%)	—	—	—
23.1 Collection period (days)	430	330	250
23.2 Collection ratio (%)	97	100	97
18.1 Average revenue W & WW (US$/m^3 water sold)	0.21	0.21	0.24
11.1 Operational cost W & WW (US$/m^3 water sold)	0.09	0.08	0.09
24.1 Operating cost coverage (ratio)	2.49	2.60	2.76
Production and consumption			
3.1 Water production (l/person/day)	120	133	130
4.1 Total water consumption (l/person/day)	78	77	73
4.7 Residential consumption (l/person/day)	38	36	35
Poverty and affordability			
19.1 Total revenues/service population GNI (% GNI per capita) (average revenues)	7	6	6
19.2 Annual bill for households consuming 6m^3 of water/month (US$/yr)	—	—	—
21.1 Ratio of industrial to residential tariff (level of cross-subsidy)	1.72	1.59	1.66

IBNET Indicator/Country: Cambodia

Latest year available	2005	2006	2007
Surface area (km²)	181,035	181,035	181,035
GNI per capita, Atlas method (current US$)	357	370	380
Total population	13,866,051	14,091,823	14,323,842
Urban population (%)	20	20	21
Total urban population	2,731,612	2,863,458	2,999,413
MDGs			
Access to improved water sources, %, 2008 (WHO and UNICEF 2010)	61	61	61
Access to improved sanitation, %, 2008 (WHO and UNICEF 2010)	29	29	29
IBNET sourced data			
Number of utilities reporting in IBNET sample	1	1	1
Population served, water (thousands)	830	910	1,068
Size of the sample: total population living in service area, water supply (thousands)	1,106	1,214	1,335
Services coverage			
1.1 Water coverage (%)	75	75	80
2.1 Sewerage coverage (%)	—	—	—
Operational efficiency			
13.2 Electrical energy costs vs. operating costs (%) (share of energy cost as % of operational expenses)	35	45	47
6.1 Nonrevenue water (%)	9	7	6
6.2 Nonrevenue water (m³/km/day)	11.60	10.30	8.20
12.3 Staff W/1,000 W population served (W/1,000 W population served)	0.60	0.60	0.50
15.1 Continuity of service (hours/day) (duration of water supply, hours)	24.00	24.00	24.00
Financial efficiency			
8.1 Water sold that is metered (%)	100	100	100
23.1 Collection period (days)	89	94	67
23.2 Collection ratio (%)	—	—	—
18.1 Average revenue W & WW (US$/m³ water sold)	0.24	0.20	0.28
11.1 Operational cost W & WW (US$/m³ water sold)	0.11	0.10	0.12
24.1 Operating cost coverage (ratio)	2.24	2.08	2.36
Production and consumption			
3.1 Water production (l/person/day)	261	271	271
4.1 Total water consumption (l/person/day)	186	197	172
4.7 Residential consumption (l/person/day)	113	118	101
Poverty and affordability			
19.1 Total revenues/service population GNI (% GNI per capita) (average revenues)	5	4	5
19.2 Annual bill for households consuming 6m³ of water/month (US$/yr)	9.68	9.41	9.69
21.1 Ratio of industrial to residential tariff (level of cross-subsidy)	1.34	1.36	1.32

IBNET Indicator/Country: Cape Verde

Latest year available	2003	2004	2005
Surface area (km^2)	4,033	4,033	4,033
GNI per Capita, Atlas method (current US$)	1,400	1,720	1,800
Total population	—	—	477,438
Urban population (%)	56	57	57
Total urban population	—	—	274,049
MDGs			
Access to improved water sources, %, 2008 (WHO and UNICEF 2010)	84	84	84
Access to improved sanitation, %, 2008 (WHO and UNICEF 2010)	54	54	54
IBNET sourced data			
Number of utilities reporting in IBNET sample	1	1	1
Population served, water (thousands)	91	101	107
Size of the sample: total population living in service area, water supply (thousands)	215	223	232
Services coverage			
1.1 Water coverage (%)	42	45	46
2.1 Sewerage coverage (%)	—	—	—
Operational efficiency			
13.2 Electrical energy costs vs. operating costs (%) (share of energy cost as % of operational expenses)	—	—	—
6.1 Nonrevenue water (%)	30	30	31
6.2 Nonrevenue water (m^3/km/day)	10.40	10.70	11.40
12.3 Staff W/1,000 W population served (W/1,000 W population served)	7.20	6.20	5.90
15.1 Continuity of service (hours/day) (duration of water supply, hours)	—	—	—
Financial efficiency			
8.1 Water sold that is metered (%)	—	—	—
23.1 Collection period (days)	—	—	—
23.2 Collection ratio (%)	—	—	—
18.1 Average revenue W & WW (US$/m^3 water sold)	3.07	3.52	3.49
11.1 Operational cost W & WW (US$/m^3 water sold)	—	—	—
24.1 Operating cost coverage (ratio)	—	—	—
Production and consumption			
3.1 Water production (l/person/day)	124	127	130
4.1 Total water consumption (l/person/day)	86	77	75
4.7 Residential consumption (l/person/day)	—	—	—
Poverty and affordability			
19.1 Total revenues/service population GNI (% GNI per capita) (average revenues)	—	—	—
19.2 Annual bill for households consuming 6m^3 of water/month (US$/yr)	—	—	—
21.1 Ratio of industrial to residential tariff (level of cross-subsidy)	—	—	—

IBNET Indicator/Country: Chile

Latest year available	2004	2005	2006
Surface area (km²)	756,102	756,102	756,102
GNI per capita, Atlas method (current US$)	5,220	5,400	5,500
Total population	—	16,297,493	16,467,256
Urban population (%)	87	88	88
Total urban population	—	14,276,604	14,471,425
MDGs			
Access to improved water sources, %, 2008 (WHO and UNICEF 2010)	96	96	96
Access to improved sanitation, %, 2008 (WHO and UNICEF 2010)	96	96	96
IBNET sourced data			
Number of utilities reporting in IBNET sample	18	18	18
Population served, water (thousands)	12,781	13,123	13,311
Size of the sample: total population living in service area, water supply (thousands)	13,215	13,155	13,340
Services coverage			
1.1 Water coverage (%)	97	100	100
2.1 Sewerage coverage (%)	92	24	99
Operational efficiency			
13.2 Electrical energy costs vs. operating costs (%) (share of energy cost as % of operational expenses)	11	6	—
6.1 Nonrevenue water (%)	33	33	33
6.2 Nonrevenue water (m³/km/day)	35.80	40.90	37.30
12.3 Staff W/1000 W population served(W/1000 W population served)	—	—	—
15.1 Continuity of service (hours/day) (duration of water supply, hours)	24.00	24.00	24.00
Financial efficiency			
8.1 Water sold that is metered (%)	98	98	98
23.1 Collection period (days)	72	80	88
23.2 Collection ratio (%)	—	122	88
18.1 Average revenue W & WW (US$/m³ water sold)	0.76	0.82	0.86
11.1 Operational cost W & WW (US$/m³ water sold)	0.29	0.58	0.62
24.1 Operating cost coverage (ratio)	2.57	1.40	1.39
Production and consumption			
3.1 Water production (l/person/day)	357	382	345
4.1 Total water consumption (l/person/day)	196	192	198
4.7 Residential consumption (l/person/day)	143	145	150
Poverty and affordability			
19.1 Total revenues/service population GNI (% GNI per capita) (average revenues)	1	1	2
19.2 Annual bill for households consuming 6m³ of water/month (US$/yr)	—	—	—
21.1 Ratio of industrial to residential tariff (level of cross-subsidy)	1.38	1.17	1.52

IBNET Indicator/Country: China

Latest year available	2007	2008	2009
Surface area (km^2)	9,640,821	9,640,821	9,640,821
GNI per capita, Atlas method (current US$)	1,870	1,700	1,870
Total population	1,317,885,000	1,327,020,000	1,333,885,000
Urban population (%)	42	43	43
Total urban population	556,147,470	541,451,260	556,147,470
MDGs			
Access to improved water sources, %, 2008 (WHO and UNICEF 2010)	89	89	89
Access to improved sanitation, %, 2008 (WHO and UNICEF 2010)	55	55	55
IBNET sourced data			
Number of utilities reporting in IBNET sample	37	37	37
Population served, water (thousands)	15,849	16,400	17,000
Size of the sample: total population living in service area, water supply (thousands)	16,627	16,819	17,600
Services coverage			
1.1 Water coverage (%)	95	93	93
2.1 Sewerage coverage (%)	—	—	—
Operational efficiency			
13.2 Electrical energy costs vs. operating costs (%) (share of energy cost as % of operational expenses)	11	13	14
6.1 Nonrevenue water (%)	21	22	21
6.2 Nonrevenue water (m^3/km/day)	54.30	50.30	50.80
12.3 Staff W/1,000 W population served (W/1,000 W population served)	1.10	1.00	1.10
15.1 Continuity of service (hours/day) (duration of water supply, hours)	24.00	24.00	24.00
Financial efficiency			
8.1 Water sold that is metered (%)	98	100	100
23.1 Collection period (days)	92	110	98
23.2 Collection ratio (%)	—	—	—
18.1 Average revenue W & WW (US$/m^3 water sold)	0.28	0.29	0.32
11.1 Operational cost W & WW (US$/m^3 water sold)	0.32	0.33	0.37
24.1 Operating cost coverage (ratio)	0.90	0.98	0.87
Production and consumption			
3.1 Water production (l/person/day)	217	200	197
4.1 Total water consumption (l/person/day)	181	167	164
4.7 Residential consumption (l/person/day)	74	72	75
Poverty and affordability			
19.1 Total revenues/service population GNI (% GNI per capita) (average revenues)	1	1	1
19.2 Annual bill for households consuming 6m^3 of water/month (US$/yr)	19.86	18.57	19.86
21.1 Ratio of industrial to residential tariff (level of cross-subsidy)	1.63	1.65	1.63

IBNET Indicator/Country: Colombia

Latest year available	2003	2004
Surface area (km²)	1,141,748	1,141,748
GNI per capita, Atlas method (current US$)	1,850	2,020
Total population	—	—
Urban population (%)	73	73
Total urban population	—	—
MDGs		
Access to improved water sources, %, 2008 (WHO and UNICEF 2010)	92	92
Access to improved sanitation, %, 2008 (WHO and UNICEF 2010)	74	74
IBNET sourced data		
Number of utilities reporting in IBNET sample	228	228
Population served, water (thousands)	22,707	23,637
Size of the sample: total population living in service area, water supply (thousands)	27,738	28,346
Services coverage		
1.1 Water coverage (%)	88	89
2.1 Sewerage coverage (%)	82	83
Operational efficiency		
13.2 Electrical energy costs vs. operating costs (%) (share of energy cost as % of operational expenses)	29	29
6.1 Nonrevenue water (%)	45	44
6.2 Nonrevenue water (m³/km/day)	91.10	87.60
12.3 Staff W/1,000 W population served (W/1,000 W population served)	0.40	0.40
15.1 Continuity of service (hours/day) (duration of water supply, hours)	—	—
Financial efficiency		
8.1 Water sold that is metered (%)	86	92
23.1 Collection period (days)	241	220
23.2 Collection ratio (%)	95	95
18.1 Average revenue W & WW (US$/m³ water sold)	0.70	0.81
11.1 Operational cost W & WW (US$/m³ water sold)	0.48	0.53
24.1 Operating cost coverage (ratio)	1.43	1.51
Production and consumption		
3.1 Water production (l/person/day)	198	196
4.1 Total water consumption (l/person/day)	146	142
4.7 Residential consumption (l/person/day)	116	112
Poverty and affordability		
19.1 Total revenues/service pop/GNI (% GNI per capita) (average revenues	2	2
19.2 Annual bill for households consuming 6m³ of water/month (US$/yr)	40.71	49.62
21.1 Ratio of industrial to residential tariff (level of cross-subsidy)	1.60	1.68

IBNET Indicator/Country: Democratic Republic of the Congo

Latest year available	2003	2004	2005
Surface area (km²)	2,344,858	2,344,858	2,344,858
GNI per capita, Atlas method (current US$)	100	110	120
Total population	—	—	59,076,752
Urban population (%)	31	32	32
Total urban population	—	—	18,963,637
MDGs			
Access to improved water sources, %, 2008 (WHO and UNICEF 2010)	46	46	46
Access to improved sanitation, %, 2008 (WHO and UNICEF 2010)	—	—	—
IBNET sourced data			
Number of utilities reporting in IBNET sample	1	1	1
Population served, water (thousands)	5,166	5,325	5,490
Size of the sample: total population living in service area, water supply (thousands)	8,468	8,730	9,000
Services coverage			
1.1 Water coverage (%)	61	61	61
2.1 Sewerage coverage (%)	24	24	24
Operational efficiency			
13.2 Electrical energy costs vs. operating costs (%) (share of energy cost as % of operational expenses)	—	—	—
6.1 Nonrevenue water (%)	44	38	35
6.2 Nonrevenue water (m³/km/day)	20.20	17.30	15.60
12.3 Staff W/1,000 W population served (W/1,000 W population served)	—	—	—
15.1 Continuity of service (hours/day) (duration of water supply, hours)	11.00	11.00	11.00
Financial efficiency			
8.1 Water sold that is metered (%)	—	—	—
23.1 Collection period (days)	1,327	2,134	1,834
23.2 Collection ratio (%)	—	—	—
18.1 Average revenue W & WW (US$/m³ water sold)	0.28	0.34	0.49
11.1 Operational cost W & WW (US$/m³ water sold)	0.77	1.04	0.76
24.1 Operating cost coverage (ratio)	0.36	0.33	0.64
Production and consumption			
3.1 Water production (l/person/day)	112	115	110
4.1 Total water consumption (l/person/day)	63	69	68
4.7 Residential consumption (l/person/day)	—	—	—
Poverty and affordability			
19.1 Total revenues/service population GNI (% GNI per capita) (average revenues)	6	8	10
19.2 Annual bill for households consuming 6m³ of water/month (US$/yr)	—	—	—
21.1 Ratio of industrial to residential tariff (level of cross-subsidy)	—	—	—

IBNET Indicator/Country: Costa Rica

Latest year available	2002	2003	2004
Surface area (km^2)	51,100	51,100	51,100
GNI per capita, Atlas method (current US$)	3,920	4,130	4,470
Total population	—	—	—
Urban population (%)	60	61	61
Total urban population	—	—	—
MDGs			
Access to improved water sources, %, 2008 (WHO and UNICEF 2010)	97	97	97
Access to improved sanitation, %, 2008 (WHO and UNICEF 2010)	95	95	95
IBNET sourced data			
Number of utilities reporting in IBNET sample	1	1	2
Population served, water (thousands)	1,689	1,832	2,242
Size of the sample: total population living in service area, water supply (thousands)	1,800	1,897	2,310
Services coverage			
1.1 Water coverage (%)	94	97	97
2.1 Sewerage coverage (%)	38	35	31
Operational efficiency			
13.2 Electrical energy costs vs. operating costs (%) (share of energy cost as % of operational expenses)	—	—	—
6.1 Nonrevenue water (%)	—	50	50
6.2 Nonrevenue water (m^3/km/day)	—	39.50	66.90
12.3 Staff W/1,000 W population served (W/1,000 W population served)	—	—	—
15.1 Continuity of service (hours/day) (duration of water supply, hours)	24.00	24.00	24.00
Financial efficiency			
8.1 Water sold that is metered (%)	—	90	23
23.1 Collection period (days)	9	22	40
23.2 Collection ratio (%)	—	—	—
18.1 Average revenue W & WW (US$/m^3 water sold)	1.32	1.58	0.56
11.1 Operational cost W & WW (US$/m^3 water sold)	1.29	0.43	0.17
24.1 Operating cost coverage (ratio)	1.02	3.63	3.27
Production and consumption			
3.1 Water production (l/person/day)	380	340	380
4.1 Total water consumption (l/person/day)	234	179	209
4.7 Residential consumption (l/person/day)	—	144	165
Poverty and affordability			
19.1 Total revenues/service population GNI (% GNI per capita) (average revenues)	—	3	1
19.2 Annual bill for households consuming 6m^3 of water/month (US$/yr)	—	—	—
21.1 Ratio of industrial to residential tariff (level of cross-subsidy)	—	—	—

IBNET Indicator/Country: Côte d'Ivoire

Latest year available	2002	2003	2004
Surface area (km^2)	322,463	322,463	322,463
GNI per capita, Atlas method (current US$)	570	630	760
Total population	—	—	—
Urban population (%)	45	45	46
Total urban population	—	—	—
MDGs			
Access to improved water sources, %, 2008 (WHO and UNICEF 2010)	80	80	80
Access to improved sanitation, %, 2008 (WHO and UNICEF 2010)	23	23	23
IBNET sourced data			
Number of utilities reporting in IBNET sample	1	1	1
Population served, water (thousands)	6,234	6,383	6,590
Size of the sample: total population living in service area, water supply (thousands)	8,180	8,426	8,678
Services coverage			
1.1 Water coverage (%)	76	76	76
2.1 Sewerage coverage (%)	26	29	26
Operational efficiency			
13.2 Electrical energy costs vs. operating costs (%) (share of energy cost as % of operational expenses)	5	4	5
6.1 Nonrevenue water (%)	19	20	21
6.2 Nonrevenue water (m^3/km/day)	6.60	7.30	7.70
12.3 Staff W/1,000 W population served (W/1,000 W population served)	0.20	0.20	0.20
15.1 Continuity of service (hours/day) (duration of water supply, hours)	24.00	24.00	24.00
Financial efficiency			
8.1 Water sold that is metered (%)	100	100	100
23.1 Collection period (days)	6	2	7
23.2 Collection ratio (%)	95	95	94
18.1 Average revenue W & WW (US$/m^3 water sold)	0.50	0.51	0.65
11.1 Operational cost W & WW (US$/m^3 water sold)	0.51	0.51	0.63
24.1 Operating cost coverage (ratio)	0.99	1.00	1.04
Production and consumption			
3.1 Water production (l/person/day)	78	76	80
4.1 Total water consumption (l/person/day)	53	53	53
4.7 Residential consumption (l/person/day)	40	40	39
Poverty and affordability			
19.1 Total revenues/service population GNI (% GNI per capita) (average revenues)	2	2	2
19.2 Annual bill for households consuming 6m^3 of water/month (US$/yr)	24.28	29.11	32.03
21.1 Ratio of industrial to residential tariff (level of cross-subsidy)	1.03	1.03	1.04

IBNET Indicator/Country: Croatia

Latest year available	2002	2003	2004
Surface area (km²)	56,594	56,594	56,594
GNI per capita, Atlas method (current US$)	4,620	5,380	6,820
Total population	—	—	—
Urban population (%)	56	56	56
Total urban population	—	—	—
MDGs			
Access to improved water sources, %, 2008 (WHO and UNICEF 2010)	99	99	99
Access to improved sanitation, %, 2008 (WHO and UNICEF 2010)	99	99	99
IBNET sourced data			
Number of utilities reporting in IBNET sample	21	21	21
Population served, water (thousands)	1,747	1,758	1,766
Size of the sample: total population living in service area, water supply (thousands)	1,894	1,899	1,903
Services coverage			
1.1 Water coverage (%)	92	93	93
2.1 Sewerage coverage (%)	75	76	76
Operational efficiency			
13.2 Electrical energy costs vs. operating costs (%) (share of energy cost as % of operational expenses)	10	10	9
6.1 Nonrevenue water (%)	17	19	19
6.2 Nonrevenue water (m³/km/day)	12.70	14.30	13.10
12.3 Staff W/1,000 W population served (W/1,000 W population served)	—	—	—
15.1 Continuity of service (hours/day) (duration of water supply, hours)	24.00	24.00	24.00
Financial efficiency			
8.1 Water sold that is metered (%)	82	82	82
23.1 Collection period (days)	114	93	114
23.2 Collection ratio (%)	71	67	60
18.1 Average revenue W & WW (US$/m³ water sold)	0.52	0.68	0.86
11.1 Operational cost W & WW (US$/m³ water sold)	0.41	0.51	0.58
24.1 Operating cost coverage (ratio)	1.27	1.33	1.47
Production and consumption			
3.1 Water production (l/person/day)	400	390	379
4.1 Total water consumption (l/person/day)	357	364	350
4.7 Residential consumption (l/person/day)	263	266	261
Poverty and affordability			
19.1 Total revenues/service population GNI (% GNI per capita) (average revenues)	1	2	2
19.2 Annual bill for households consuming 6m³ of water/month (US$/yr)	—	—	—
21.1 Ratio of industrial to residential tariff (level of cross-subsidy)	4.00	9.92	4.38

IBNET Indicator/Country: Czech Republic

Latest year available	2003	2004	2005
Surface area (km^2)	78,867	78,867	78,867
GNI per capita, Atlas method (current US$)	7,160	9,130	10,000
Total population	—	—	10,235,828
Urban population (%)	74	74	74
Total urban population	—	—	7,523,334
MDGs			
Access to improved water sources, %, 2008 (WHO and UNICEF 2010)	100	100	100
Access to improved sanitation, %, 2008 (WHO and UNICEF 2010)	98	98	98
IBNET sourced data			
Number of utilities reporting in IBNET sample	20	20	20
Population served, water (thousands)	5,195	5,214	5,216
Size of the sample: total population living in service area, water supply (thousands)	5,699	5,732	5,750
Services coverage			
1.1 Water coverage (%)	91	91	91
2.1 Sewerage coverage (%)	76	77	77
Operational efficiency			
13.2 Electrical energy costs vs. operating costs (%) (share of energy cost as % of operational expenses)	6	6	3
6.1 Nonrevenue water (%)	22	20	20
6.2 Nonrevenue water (m^3/km/day)	10.50	8.60	8.50
12.3 Staff W/1,000 W population served (W/1,000 W population served)	1.00	1.00	0.80
15.1 Continuity of service (hours/day) (duration of water supply, hours)	24.00	24.00	24.00
Financial efficiency			
8.1 Water sold that is metered (%)	99	99	99
23.1 Collection period (days)	121	133	168
23.2 Collection ratio (%)	97	98	98
18.1 Average revenue W & WW (US$/m^3 water sold)	1.17	1.37	1.54
11.1 Operational cost W & WW (US$/m^3 water sold)	0.90	1.03	1.06
24.1 Operating cost coverage (ratio)	1.30	1.30	1.33
Production and consumption			
3.1 Water production (l/person/day)	242	239	230
4.1 Total water consumption (l/person/day)	201	198	191
4.7 Residential consumption (l/person/day)	107	106	102
Poverty and affordability			
19.1 Total revenues/service population GNI (% GNI per capita) (average revenues)	1	1	1
19.2 Annual bill for households consuming 6m^3 of water/month (US$/yr)	48.62	52.02	58.03
21.1 Ratio of industrial to residential tariff (level of cross-subsidy)	1.01	0.96	1.08

IBNET Indicator/Country: Ecuador

Latest year available	2003	2004	2005
Surface area (km^2)	256,369	256,369	256,369
GNI per capita, Atlas method (current US$)	1,850	2,210	2,200
Total population	—	—	13,062,507
Urban population (%)	62	63	64
Total urban population	—	—	8,307,754
MDGs			
Access to improved water sources, %, 2008 (WHO and UNICEF 2010)	94	94	94
Access to improved sanitation, %, 2008 (WHO and UNICEF 2010)	92	92	92
IBNET sourced data			
Number of utilities reporting in IBNET sample	1	1	1
Population served, water (thousands)	1,366	1,463	1,494
Size of the sample: total population living in service area, water supply (thousands)	2,172	2,164	2,207
Services coverage			
1.1 Water coverage (%)	63	68	68
2.1 Sewerage coverage (%)	28	32	34
Operational efficiency			
13.2 Electrical energy costs vs. operating costs (%) (share of energy cost as % of operational expenses)	36	36	13
6.1 Nonrevenue water (%)	73	74	71
6.2 Nonrevenue water (m^3/km/day)	190.30	214.90	205.80
12.3 Staff W/1,000 W population served (W/1,000 W population served)	—	—	0.20
15.1 Continuity of service (hours/day) (duration of water supply, hours)	24.00	24.00	24.00
Financial efficiency			
8.1 Water sold that is metered (%)	59	83	80
23.1 Collection period (days)	148	111	151
23.2 Collection ratio (%)	—	—	95
18.1 Average revenue W & WW (US$/m^3 water sold)	0.70	0.74	0.72
11.1 Operational cost W & WW (US$/m^3 water sold)	0.29	0.26	0.70
24.1 Operating cost coverage (ratio)	2.44	2.85	1.04
Production and consumption			
3.1 Water production (l/person/day)	332	338	337
4.1 Total water consumption (l/person/day)	183	163	178
4.7 Residential consumption (l/person/day)	94	92	101
Poverty and affordability			
19.1 Total revenues/service population GNI (% GNI per capita) (average revenues)	3	2	2
19.2 Annual bill for households consuming 6m^3 of water/month (US$/yr)	—	—	—
21.1 Ratio of industrial to residential tariff (level of cross-subsidy)	—	—	—

IBNET Indicator/Country: El Salvador

Latest year available	2006
Surface area (km^2)	21,041
GNI per capita, Atlas method (current US$)	2,680
Total population	6,081,703
Urban population (%)	60
Total urban population	3,655,104
MDGs	
Access to improved water sources, %, 2008 (WHO and UNICEF 2010)	87
Access to improved sanitation, %, 2008 (WHO and UNICEF 2010)	87
IBNET sourced data	
Number of utilities reporting in IBNET sample	1
Population served, water (thousands)	3,951
Size of the sample: total population living in service area, water supply (thousands)	5,382
Services coverage	
1.1 Water coverage (%)	73
2.1 Sewerage coverage (%)	39
Operational efficiency	
13.2 Electrical energy costs vs. operating costs (%) (share of energy cost as % of operational expenses)	—
6.1 Nonrevenue water (%)	34
6.2 Nonrevenue water (m^3/km/day)	74.70
12.3 Staff W/1,000 W population served (W/1,000 W population served)	0.50
15.1 Continuity of service (hours/day) (duration of water supply, hours)	24.00
Financial efficiency	
8.1 Water sold that is metered (%)	65
23.1 Collection period (days)	93
23.2 Collection ratio (%)	104
18.1 Average revenue W & WW (US$/m^3 water sold)	0.04
11.1 Operational cost W & WW (US$/m^3 water sold)	0.03
24.1 Operating cost coverage (ratio)	1.17
Production and consumption	
3.1 Water production (l/person/day)	348
4.1 Total water consumption (l/person/day)	158
4.7 Residential consumption (l/person/day)	122
Poverty and affordability	
19.1 Total revenues/service population GNI (% GNI per capita) (average revenues)	—
19.2 Annual bill for households consuming 6m^3 of water/month (US$/yr)	—
21.1 Ratio of industrial to residential tariff (level of cross-subsidy)	—

IBNET Indicator/Country: Ethiopia

Latest year available	2004	2005	2006
Surface area (km^2)	1,104,300	1,104,300	1,104,300
GNI per capita, Atlas method (current US$)	115	120	135
Total population	—	74,660,901	76,627,697
Urban population (%)	16	16	16
Total urban population	—	12,020,405	12,566,942
MDGs			
Access to improved water sources, %, 2008 (WHO and UNICEF 2010)	38	38	38
Access to improved sanitation, %, 2008 (WHO and UNICEF 2010)	12	12	12
IBNET sourced data			
Number of utilities reporting in IBNET sample	6	6	6
Population served, water (thousands)	3,313	3,515	3,648
Size of the sample: total population living in service area, water supply (thousands)	3,838	4,006	4,135
Services coverage			
1.1 Water coverage (%)	86	88	88
2.1 Sewerage coverage (%)	—	—	—
Operational efficiency			
13.2 Electrical energy costs vs. operating costs (%) (share of energy cost as % of operational expenses)	—	—	—
6.1 Nonrevenue water (%)	35	35	33
6.2 Nonrevenue water (m^3/km/day)	32.90	33.30	31.70
12.3 Staff W/1,000 W population served (W/1,000 W population served)	0.50	0.50	0.50
15.1 Continuity of service (hours/day) (duration of water supply, hours)	22.67	22.67	22.67
Financial efficiency			
8.1 Water sold that is metered (%)	—	—	100
23.1 Collection period (days)	54	61	87
23.2 Collection ratio (%)	27	29	36
18.1 Average revenue W & WW (US$/m^3 water sold)	1.04	1.02	0.98
11.1 Operational cost W & WW (US$/m^3 water sold)	0.33	0.27	0.25
24.1 Operating cost coverage (ratio)	3.30	3.98	4.18
Production and consumption			
3.1 Water production (l/person/day)	85	93	96
4.1 Total water consumption (l/person/day)	48	47	48
4.7 Residential consumption (l/person/day)	25	25	26
Poverty and affordability			
19.1 Total revenues/service population GNI (% GNI per capita) (average revenues)	—	—	14
19.2 Annual bill for households consuming 6m^3 of water/month (US$/yr)	—	—	12.93
21.1 Ratio of industrial to residential tariff (level of cross-subsidy)	—	—	1.01

IBNET Indicator/Country: Gabon

Latest year available	2003	2004	2005
Surface area (km^2)	267,668	267,668	267,668
GNI per capita, Atlas method (current US$)	3,340	4,080	4,200
Total population	—	—	1,369,229
Urban population (%)	82	83	84
Total urban population	—	—	1,144,675
MDGs			
Access to improved water sources, %, 2008 (WHO and UNICEF 2010)	87	87	87
Access to improved sanitation, %, 2008 (WHO and UNICEF 2010)	33	33	33
IBNET sourced data			
Number of utilities reporting in IBNET sample	1	1	1
Population served, water (thousands)	629	676	728
Size of the sample: total population living in service area, water supply (thousands)	1,065	1,093	1,121
Services coverage			
1.1 Water coverage (%)	59	62	65
2.1 Sewerage coverage (%)	—	—	—
Operational efficiency			
13.2 Electrical energy costs vs. operating costs (%) (share of energy cost as % of operational expenses)	—	—	—
6.1 Nonrevenue water (%)	17	16	18
6.2 Nonrevenue water (m^3/km/day)	17.60	17.30	19.10
12.3 Staff W/1,000 W population served (W/1,000 W population served)	0.90	0.80	0.80
15.1 Continuity of service (hours/day) (duration of water supply, hours)	24.00	24.00	24.00
Financial efficiency			
8.1 Water sold that is metered (%)	—	—	—
23.1 Collection period (days)	97	125	114
23.2 Collection ratio (%)	—	—	—
18.1 Average revenue W & WW (US$/m^3 water sold)	0.36	0.40	0.41
11.1 Operational cost W & WW (US$/m^3 water sold)	0.31	0.37	0.40
24.1 Operating cost coverage (ratio)	1.16	1.09	1.01
Production and consumption			
3.1 Water production (l/person/day)	260	263	267
4.1 Total water consumption (l/person/day)	216	214	207
4.7 Residential consumption (l/person/day)	—	—	—
Poverty and affordability			
19.1 Total revenues/service population GNI (% GNI per capita) (average revenues)	—	—	—
19.2 Annual bill for households consuming 6m^3 of water/month (US$/yr)	—	—	—
21.1 Ratio of industrial to residential tariff (level of cross-subsidy)	—	—	—

IBNET Indicator/Country: The Gambia

Latest year available	2005
Surface area (km²)	11,295
GNI per capita, Atlas method (current US$)	300
Total population	1,526,138
Urban population (%)	54
Total urban population	822,588
MDGs	
Access to improved water sources, %, 2008 (WHO and UNICEF 2010)	92
Access to improved sanitation, %, 2008 (WHO and UNICEF 2010)	67
IBNET sourced data	
Number of utilities reporting in IBNET sample	1
Population served, water (thousands)	626
Size of the sample: total population living in service area, water supply (thousands)	821
Services coverage	
1.1 Water coverage (%)	76
2.1 Sewerage coverage (%)	5
Operational efficiency	
13.2 Electrical energy costs vs. operating costs (%) (share of energy cost as % of operational expenses)	—
6.1 Nonrevenue water (%)	17
6.2 Nonrevenue water (m³/km/day)	25.00
12.3 Staff W/1,000 W population served (W/1,000 W population served)	—
15.1 Continuity of service (hours/day) (duration of water supply, hours)	—
Financial efficiency	
8.1 Water sold that is metered (%)	—
23.1 Collection period (days)	—
23.2 Collection ratio (%)	94
18.1 Average revenue W & WW (US$/m³ water sold)	0.26
11.1 Operational cost W & WW (US$/m³ water sold)	0.33
24.1 Operating cost coverage (ratio)	0.79
Production and consumption	
3.1 Water production (l/person/day)	86
4.1 Total water consumption (l/person/day)	57
4.7 Residential consumption (l/person/day)	—
Poverty and affordability	
19.1 Total revenues/service population GNI (% GNI per capita) (average revenues)	—
19.2 Annual bill for households consuming 6m³ of water/month (US$/yr)	—
21.1 Ratio of industrial to residential tariff (level of cross-subsidy)	—

IBNET Indicator/Country: Georgia

Latest year available	2006	2007	2008
Surface area (km^2)	69,700	69,700	69,700
GNI per capita, Atlas method (current US$)	1,150	1,175	1,200
Total population	4,410,860	4,357,857	4,307,011
Urban population (%)	53	53	53
Total urban population	2,319,230	2,294,847	2,271,518
MDGs			
Access to improved water sources, %, 2008 (WHO and UNICEF 2010)	98	98	98
Access to improved sanitation, %, 2008 (WHO and UNICEF 2010)	95	95	95
IBNET sourced data			
Number of utilities reporting in IBNET sample	17	17	16
Population served, water (thousands)	1,230	1,242	1,260
Size of the sample: total population living in service area, water supply (thousands)	1,301	1,303	1,318
Services coverage			
1.1 Water coverage (%)	95	95	96
2.1 Sewerage coverage (%)	84	83	82
Operational efficiency			
13.2 Electrical energy costs vs. operating costs (%) (share of energy cost as % of operational expenses)	14	15	18
6.1 Nonrevenue water (%)	43	43	43
6.2 Nonrevenue water (m^3/km/day)	126.10	131.00	128.90
12.3 Staff W/1,000 W population served (W/1,000 W population served)	2.60	2.50	2.60
15.1 Continuity of service (hours/day) (duration of water supply, hours)	14.07	14.71	14.71
Financial efficiency			
8.1 Water sold that is metered (%)	8	8	8
23.1 Collection period (days)	207	152	92
23.2 Collection ratio (%)	105	98	114
18.1 Average revenue W & WW (US$/m^3 water sold)	0.09	0.13	0.14
11.1 Operational cost W & WW (US$/m^3 water sold)	0.18	0.14	0.14
24.1 Operating cost coverage (ratio)	0.50	0.95	0.98
Production and consumption			
3.1 Water production (l/person/day)	1,241	1,262	1,362
4.1 Total water consumption (l/person/day)	685	702	701
4.7 Residential consumption (l/person/day)	603	619	616
Poverty and affordability			
19.1 Total revenues/service population GNI (% GNI per capita) (average revenues)	2	3	3
19.2 Annual bill for households consuming 6m^3 of water/month (US$/yr)	4.94	7.49	7.77
21.1 Ratio of industrial to residential tariff (level of cross-subsidy)	36.38	47.78	45.44

IBNET Indicator/Country: Ghana

Latest year available	2003	2004	2005
Surface area (km^2)	238,539	238,539	238,539
GNI per capita, Atlas method (current US$)	310	380	400
Total population	—	—	21,915,168
Urban population (%)	46	47	48
Total urban population	—	—	10,475,450
MDGs			
Access to improved water sources, %, 2008 (WHO and UNICEF 2010)	82	82	82
Access to improved sanitation, %, 2008 (WHO and UNICEF 2010)	13	13	13
IBNET sourced data			
Number of utilities reporting in IBNET sample	1	1	1
Population served, water (thousands)	4,974	5,108	5,246
Size of the sample: total population living in service area, water supply (thousands)	8,154	8,374	8,600
Services coverage			
1.1 Water coverage (%)	61	61	61
2.1 Sewerage coverage (%)	24	24	24
Operational efficiency			
13.2 Electrical energy costs vs. operating costs (%) (share of energy cost as % of operational expenses)	—	—	—
6.1 Nonrevenue water (%)	57	53	53
6.2 Nonrevenue water (m^3/km/day)	51.50	46.60	39.50
12.3 Staff W/1,000 W population served (W/1,000 W population served)	—	—	—
15.1 Continuity of service (hours/day) (duration of water supply, hours)	11.00	11.00	11.00
Financial efficiency			
8.1 Water sold that is metered (%)	—	—	—
23.1 Collection period (days)	—	—	—
23.2 Collection ratio (%)	—	—	—
18.1 Average revenue W & WW (US$/m^3 water sold)	0.52	0.56	0.60
11.1 Operational cost W & WW (US$/m^3 water sold)	0.44	0.44	0.53
24.1 Operating cost coverage (ratio)	1.20	1.29	1.13
Production and consumption			
3.1 Water production (l/person/day)	105	108	110
4.1 Total water consumption (l/person/day)	49	52	51
4.7 Residential consumption (l/person/day)	—	—	—
Poverty and affordability			
19.1 Total revenues/service population GNI (% GNI per capita) (average revenues)	—	—	—
19.2 Annual bill for households consuming 6m^3 of water/month (US$/yr)	—	—	—
21.1 Ratio of industrial to residential tariff (level of cross-subsidy)	—	—	—

IBNET Indicator/Country: Guinea

Latest year available	2004	2005	2006
Surface area (km^2)	245,857	245,857	245,857
GNI per capita, Atlas method (current US$)	410	420	440
Total population	—	9,220,768	9,411,881
Urban population (%)	33	33	33
Total urban population	—	3,042,853	3,151,098
MDGs			
Access to improved water sources, %, 2008 (WHO and UNICEF 2010)	71	71	71
Access to improved sanitation, %, 2008 (WHO and UNICEF 2010)	19	19	19
IBNET sourced data			
Number of utilities reporting in IBNET sample	1	1	1
Population served, water (thousands)	1,800	1,900	2,000
Size of the sample: total population living in service area, water supply (thousands)	2,400	2,500	2,600
Services coverage			
1.1 Water coverage (%)	75	76	77
2.1 Sewerage coverage (%)	—	—	—
Operational efficiency			
13.2 Electrical energy costs vs. operating costs (%) (share of energy cost as % of operational expenses)	—	—	—
6.1 Nonrevenue water (%)	45	50	48
6.2 Nonrevenue water (m^3/km/day)	24.90	27.90	28.80
12.3 Staff W/1,000 W population served (W/1,000 W population served)	—	—	—
15.1 Continuity of service (hours/day) (duration of water supply, hours)	8.00	8.00	8.00
Financial efficiency			
8.1 Water sold that is metered (%)	100	100	100
23.1 Collection period (days)	616	490	484
23.2 Collection ratio (%)	67	60	59
18.1 Average revenue W & WW (US$/m^3 water sold)	0.40	0.59	0.69
11.1 Operational cost W & WW (US$/m^3 water sold)	0.82	1.24	1.17
24.1 Operating cost coverage (ratio)	0.49	0.48	0.59
Production and consumption			
3.1 Water production (l/person/day)	50	44	43
4.1 Total water consumption (l/person/day)	25	22	23
4.7 Residential consumption (l/person/day)	—	—	—
Poverty and affordability			
19.1 Total revenues/service population GNI (% GNI per capita) (average revenues)	1	1	1
19.2 Annual bill for households consuming 6m^3 of water/month (US$/yr)	—	—	—
21.1 Ratio of industrial to residential tariff (level of cross-subsidy)	—	—	—

IBNET Indicator/Country: Honduras

Latest year available	2002	2003	2004
Surface area (km^2)	112,492	112,492	112,492
GNI per capita, Atlas method (current US$)	920	970	1,040
Total population	—	—	—
Urban population (%)	45	46	46
Total urban population	—	—	—
MDGs			
Access to improved water sources, %, 2008 (WHO and UNICEF 2010)	86	86	86
Access to improved sanitation, %, 2008 (WHO and UNICEF 2010)	71	71	71
IBNET sourced data			
Number of utilities reporting in IBNET sample	1	1	1
Population served, water (thousands)	43	44	60
Size of the sample: total population living in service area, water supply (thousands)	54	54	100
Services coverage			
1.1 Water coverage (%)	80	81	60
2.1 Sewerage coverage (%)	—	—	—
Operational efficiency			
13.2 Electrical energy costs vs. operating costs (%) (share of energy cost as % of operational expenses)	—	—	—
6.1 Nonrevenue water (%)	60	65	23
6.2 Nonrevenue water (m^3/km/day)	458.60	328.40	88.50
12.3 Staff W/1,000 W population served (W/1,000 W population served)	—	—	—
15.1 Continuity of service (hours/day) (duration of water supply, hours)	24.00	24.00	24.00
Financial efficiency			
8.1 Water sold that is metered (%)	—	100	51
23.1 Collection period (days)	—	—	—
23.2 Collection ratio (%)	—	—	—
18.1 Average revenue W & WW (US$/m^3 water sold)	—	—	—
11.1 Operational cost W & WW (US$/m^3 water sold)	—	—	—
24.1 Operating cost coverage (ratio)	—	—	—
Production and consumption			
3.1 Water production (l/person/day)	319	411	440
4.1 Total water consumption (l/person/day)	167	244	389
4.7 Residential consumption (l/person/day)	—	188	137
Poverty and affordability			
19.1 Total revenues/service population GNI (% GNI per capita) (average revenues)	—	—	—
19.2 Annual bill for households consuming 6m^3 of water/month (US$/yr)	—	—	—
21.1 Ratio of industrial to residential tariff (level of cross-subsidy)	—	—	—

IBNET Indicator/Country: Hungary

Latest year available	2005	2006	2007
Surface area (km²)	93,028	93,028	93,028
GNI per capita, Atlas method (current US$)	8,400	8,500	8,750
Total population	10,087,050	10,071,370	10,055,780
Urban population (%)	66	67	67
Total urban population	6,687,714	6,717,604	6,747,428
MDGs			
Access to improved water sources, %, 2008 (WHO and UNICEF 2010)	100	100	100
Access to improved sanitation, %, 2008 (WHO and UNICEF 2010)	100	100	100
IBNET sourced data			
Number of utilities reporting in IBNET sample	24	20	20
Population served, water (thousands)	5,379	4,853	4,853
Size of the sample: total population living in service area, water supply (thousands)	5,440	4,902	4,902
Services coverage			
1.1 Water coverage (%)	99	99	99
2.1 Sewerage coverage (%)	42	70	70
Operational efficiency			
13.2 Electrical energy costs vs. operating costs (%) (share of energy cost as % of operational expenses)	9	10	11
6.1 Nonrevenue water (%)	20	34	32
6.2 Nonrevenue water (m³/km/day)	8.00	15.20	14.20
12.3 Staff W/1,000 W population served (W/1,000 W population served)	1.10	0.90	0.90
15.1 Continuity of service (hours/day) (duration of water supply, hours)	23.00	24.00	24.00
Financial efficiency			
8.1 Water sold that is metered (%)	96	100	100
23.1 Collection period (days)	33	45	49
23.2 Collection ratio (%)	100	94	101
18.1 Average revenue W & WW (US$/m³ water sold)	1.20	1.37	1.64
11.1 Operational cost W & WW (US$/m³ water sold)	0.93	1.33	1.51
24.1 Operating cost coverage (ratio)	1.18	1.03	1.09
Production and consumption			
3.1 Water production (l/person/day)	436	426	413
4.1 Total water consumption (l/person/day)	178	159	158
4.7 Residential consumption (l/person/day)	114	109	110
Poverty and affordability			
19.1 Total revenues/service population GNI (% GNI per capita) (average revenues)	1	1	1
19.2 Annual bill for households consuming 6m³ of water/month (US$/yr)	73.18	77.95	99.80
21.1 Ratio of industrial to residential tariff (level of cross-subsidy)	1.25	1.32	1.22

IBNET Indicator/Country: India

Latest year available	2005	2009
Surface area (km^2)	3,287,240	3,287,240
GNI per capita, Atlas method (current US$)	660	1,134
Total population	1,094,583,000	1,180,166,000
Urban population (%)	29	33
Total urban population	314,145,321	389,454,780
MDGs		
Access to improved water sources, %, 2008 (WHO and UNICEF 2010)	88	88
Access to improved sanitation, %, 2008 (WHO and UNICEF 2010)	31	31
IBNET sourced data		
Number of utilities reporting in IBNET sample	25	27
Population served, water (thousands)	23,458	57,399
Size of the sample: total population living in service area, water supply (thousands)	26,332	57,398
Services coverage		
1.1 Water coverage (%)	89	100
2.1 Sewerage coverage (%)	68	23
Operational efficiency		
13.2 Electrical energy costs vs. operating costs (%) (share of energy cost as % of operational expenses)	41	38
6.1 Nonrevenue water (%)	33	41.00
6.2 Nonrevenue water (m^3/km/day)	84.20	119
12.3 Staff W/1,000 W population served (W/1,000 W population served)	0.60	0.60
15.1 Continuity of service (hours/day) (duration of water supply, hours)	4.41	5.20
Financial efficiency		
8.1 Water sold that is metered (%)	58	39
23.1 Collection period (days)	256	200
23.2 Collection ratio (%)	68	82
18.1 Average revenue W & WW (US$/m^3 water sold)	0.23	0.15
11.1 Operational cost W & WW (US$/m^3 water sold)	0.28	0.28
24.1 Operating cost coverage (ratio)	0.83	0.55
Production and consumption		
3.1 Water production (l/person/day)	170	193
4.1 Total water consumption (l/person/day)	134	114
4.7 Residential consumption (l/person/day)	—	83
Poverty and affordability		
19.1 Total revenues/service population GNI (% GNI per capita) (average revenues)	—	2
19.2 Annual bill for households consuming 6m^3 of water/month (US$/yr)	—	6.09
21.1 Ratio of industrial to residential tariff (level of cross-subsidy)	—	9.32

IBNET Indicator/Country: Indonesia

Latest year available	2002	2003	2004
Surface area (km^2)	1,860,360	1,860,360	1,860,360
GNI per capita, Atlas method (current US$)	830	940	1,140
Total population	—	—	—
Urban population (%)	44	46	47
Total urban population	—	—	—
MDGs			
Access to improved water sources, %, 2008 (WHO and UNICEF 2010)	80	80	80
Access to improved sanitation, %, 2008 (WHO and UNICEF 2010)	52	52	52
IBNET sourced data			
Number of utilities reporting in IBNET sample	14	14	7
Population served, water (thousands)	4,729	5,308	1,952
Size of the sample: total population living in service area, water supply (thousands)	10,530	10,874	2,571
Services coverage			
1.1 Water coverage (%)	45	49	76
2.1 Sewerage coverage (%)	11	12	15
Operational efficiency			
13.2 Electrical energy costs vs. operating costs (%) (share of energy cost as % of operational expenses)	17	17	17
6.1 Nonrevenue water (%)	30	30	30
6.2 Nonrevenue water (m^3/km/day)	36.20	37.00	27.70
12.3 Staff W/1,000 W population served (W/1,000 W population served)	1.10	1.00	1.00
15.1 Continuity of service (hours/day) (duration of water supply, hours)	19.14	19.79	19.86
Financial efficiency			
8.1 Water sold that is metered (%)	100	100	100
23.1 Collection period (days)	46	55	56
23.2 Collection ratio (%)	111	110	110
18.1 Average revenue W & WW (US$/m^3 water sold)	0.14	0.18	0.20
11.1 Operational cost W & WW (US$/m^3 water sold)	0.12	0.15	0.15
24.1 Operating cost coverage (ratio)	1.22	1.21	1.39
Production and consumption			
3.1 Water production (l/person/day)	373	394	133
4.1 Total water consumption (l/person/day)	151	142	130
4.7 Residential consumption (l/person/day)	133	123	117
Poverty and affordability			
19.1 Total revenues/service population GNI (% GNI per capita) (average revenues)	—	—	—
19.2 Annual bill for households consuming 6m^3 of water/month (US$/yr)	—	—	—
21.1 Ratio of industrial to residential tariff (level of cross-subsidy)	—	—	—

IBNET Indicator/Country: Kazakhstan

Latest year available	2005	2006	2007
Surface area (km²)	2,724,900	2,724,900	2,724,900
GNI per capita, Atlas method (current US$)	2,600	2,900	3,000
Total population	15,147,000	15,308,100	15,484,200
Urban population (%)	57	57	58
Total urban population	8,648,937	8,783,788	8,928,190
MDGs			
Access to improved water sources, %, 2008 (WHO and UNICEF 2010)	95	95	95
Access to improved sanitation, %, 2008 (WHO and UNICEF 2010)	97	97	97
IBNET sourced data			
Number of utilities reporting in IBNET sample	19	22	24
Population served, water (thousands)	4,382	5,049	5,353
Size of the sample: total population living in service area, water supply (thousands)	5,145	5,947	6,261
Services coverage			
1.1 Water coverage (%)	85	85	86
2.1 Sewerage coverage (%)	64	64	65
Operational efficiency			
13.2 Electrical energy costs vs. operating costs (%) (share of energy cost as % of operational expenses)	20	19	21
6.1 Nonrevenue water (%)	35	35	32
6.2 Nonrevenue water (m³/km/day)	66.10	65.60	59.20
12.3 Staff W/1,000 W population served (W/1,000 W population served)	1.20	1.40	1.30
15.1 Continuity of service (hours/day) (duration of water supply, hours)	24.00	24.00	24.00
Financial efficiency			
8.1 Water sold that is metered (%)	60	55	55
23.1 Collection period (days)	82	90	94
23.2 Collection ratio (%)	103	94	95
18.1 Average revenue W & WW (US$/m³ water sold)	0.17	0.22	0.24
11.1 Operational cost W & WW (US$/m³ water sold)	0.20	0.24	0.27
24.1 Operating cost coverage (ratio)	0.88	0.92	0.87
Production and consumption			
3.1 Water production (l/person/day)	391	413	353
4.1 Total water consumption (l/person/day)	281	289	295
4.7 Residential consumption (l/person/day)	127	123	122
Poverty and affordability			
19.1 Total revenues/service population GNI (% GNI per capita) (average revenues)	1	1	1
19.2 Annual bill for households consuming 6m³ of water/month (US$/yr)	22.84	27.05	27.91
21.1 Ratio of industrial to residential tariff (level of cross-subsidy)	0.96	1.02	0.99

IBNET Indicator/Country: Kenya

Latest year available	2004	2005	2006
Surface area (km²)	580,367	580,367	580,367
GNI per capita, Atlas method (current US$)	480	500	520
Total population	—	35,816,784	36,771,613
Urban population (%)	21	21	21
Total urban population	—	7,414,074	7,722,039
MDGs			
Access to improved water sources, %, 2008 (WHO and UNICEF 2010)	59	59	59
Access to improved sanitation, %, 2008 (WHO and UNICEF 2010)	31	31	31
IBNET sourced data			
Number of utilities reporting in IBNET sample	7	7	7
Population served, water (thousands)	3,537	3,736	3,952
Size of the sample: total population living in service area, water supply (thousands)	5,493	5,684	5,947
Services coverage			
1.1 Water coverage (%)	66	67	66
2.1 Sewerage coverage (%)	15	17	32
Operational efficiency			
13.2 Electrical energy costs vs. operating costs (%) (share of energy cost as % of operational expenses)	6	12	12
6.1 Nonrevenue water (%)	52	36	49
6.2 Nonrevenue water (m³/km/day)	23.80	28.60	119.40
12.3 Staff W/1,000 W population served (W/1,000 W population served)	0.30	0.50	0.50
15.1 Continuity of service (hours/day) (duration of water supply, hours)	14.17	14.83	15.83
Financial efficiency			
8.1 Water sold that is metered (%)	—	—	88
23.1 Collection period (days)	60	41	85
23.2 Collection ratio (%)	176	137	113
18.1 Average revenue W & WW (US$/m³ water sold)	0.96	0.32	0.48
11.1 Operational cost W & WW (US$/m³ water sold)	0.31	0.22	0.25
24.1 Operating cost coverage (ratio)	0.74	1.45	1.91
Production and consumption			
3.1 Water production (l/person/day)	212	161	149
4.1 Total water consumption (l/person/day)	110	96	77
4.7 Residential consumption (l/person/day)	—	—	—
Poverty and affordability			
19.1 Total revenues/service population GNI (% GNI per capita) (average revenues)	2	1	3
19.2 Annual bill for households consuming 6m³ of water/month (US$/yr)	—	—	—
21.1 Ratio of industrial to residential tariff (level of cross-subsidy)	—	—	—

IBNET Indicator/Country: Kyrgyz Republic

Latest year available	2004	2005	2006
Surface area (km^2)	199,951	199,951	199,951
GNI per capita, Atlas method (current US$)	400	410	400
Total population	—	5,143,500	5,192,100
Urban population (%)	36	36	36
Total urban population	—	1,841,373	1,867,079
MDGs			
Access to improved water sources, %, 2008 (WHO and UNICEF 2010)	90	90	90
Access to improved sanitation, %, 2008 (WHO and UNICEF 2010)	93	93	93
IBNET sourced data			
Number of utilities reporting in IBNET sample	9	9	9
Population served, water (thousands)	347	361	376
Size of the sample: total population living in service area, water supply (thousands)	625	656	669
Services coverage			
1.1 Water coverage (%)	55	55	56
2.1 Sewerage coverage (%)	16	15	15
Operational efficiency			
13.2 Electrical energy costs vs. operating costs (%) (share of energy cost as % of operational expenses)	37	37	36
6.1 Nonrevenue water (%)	70	69	70
6.2 Nonrevenue water (m^3/km/day)	91.90	86.30	83.20
12.3 Staff W/1,000 W population served (W/1,000 W population served)	1.90	1.80	1.80
15.1 Continuity of service (hours/day) (duration of water supply, hours)	23.33	23.56	23.56
Financial efficiency			
8.1 Water sold that is metered (%)	11	8	10
23.1 Collection period (days)	299	350	514
23.2 Collection ratio (%)	62	60	58
18.1 Average revenue W & WW (US$/m^3 water sold)	0.12	0.12	0.11
11.1 Operational cost W & WW (US$/m^3 water sold)	0.10	0.11	0.11
24.1 Operating cost coverage (ratio)	1.11	1.15	1.03
Production and consumption			
3.1 Water production (l/person/day)	262	264	262
4.1 Total water consumption (l/person/day)	147	150	137
4.7 Residential consumption (l/person/day)	75	78	64
Poverty and affordability			
19.1 Total revenues/service population GNI (% GNI per capita) (average revenues)	2	2	1
19.2 Annual bill for households consuming 6m^3 of water/month (US$/yr)	5.16	5.36	6.60
21.1 Ratio of industrial to residential tariff (level of cross-subsidy)	2.35	2.16	1.97

IBNET Indicator/Country: Lao People's Democratic Republic

Latest year available	2006	2007	2008
Surface area (km^2)	236,800	236,800	236,800
GNI per capita, Atlas method (current US$)	420	440	460
Total population	5,983,451	6,092,332	6,205,341
Urban population (%)	29	30	31
Total urban population	1,611,087	1,708,874	1,810,641
MDGs			
Access to improved water sources, %, 2008 (WHO and UNICEF 2010)	57	57	57
Access to improved sanitation, %, 2008 (WHO and UNICEF 2010)	53	53	53
IBNET sourced data			
Number of utilities reporting in IBNET sample	2	10	2
Population served, water (thousands)	321	596	57
Size of the sample: total population living in service area, water supply (thousands)	658	1,055	66
Services coverage			
1.1 Water coverage (%)	49	56	87
2.1 Sewerage coverage (%)	—	—	—
Operational efficiency			
13.2 Electrical energy costs vs. operating costs (%) (share of energy cost as % of operational expenses)	15	11	16
6.1 Nonrevenue water (%)	27	26	21
6.2 Nonrevenue water (m^3/km/day)	46.90	25.70	45.80
12.3 Staff W/1,000 W population served (W/1,000 W population served)	1.60	1.60	2.20
15.1 Continuity of service (hours/day) (duration of water supply, hours)	24.00	24.00	24.00
Financial efficiency			
8.1 Water sold that is metered (%)	100	100	100
23.1 Collection period (days)	77	101	49
23.2 Collection ratio (%)	—	—	—
18.1 Average revenue W & WW (US$/m^3 water sold)	0.09	0.13	0.15
11.1 Operational cost W & WW (US$/m^3 water sold)	0.15	0.24	0.14
24.1 Operating cost coverage (ratio)	0.58	0.54	1.07
Production and consumption			
3.1 Water production (l/person/day)	344	373	371
4.1 Total water consumption (l/person/day)	274	249	250
4.7 Residential consumption (l/person/day)	223	197	154
Poverty and affordability			
19.1 Total revenues/service population GNI (% GNI per capita) (average revenues)	2	3	3
19.2 Annual bill for households consuming 6m^3 of water/month (US$/yr)	—	—	—
21.1 Ratio of industrial to residential tariff (level of cross-subsidy)	4.02	3.45	2.12

IBNET Indicator/Country: Lesotho

Latest year available	2006	2007	2008
Surface area (km²)	30,355	30,355	30,355
GNI per capita, Atlas method (current US$)	800	810	800
Total population	2,013,620	2,031,676	2,049,429
Urban population (%)	24	25	25
Total urban population	483,672	502,637	521,785
MDGs			
Access to improved water sources, %, 2008 (WHO and UNICEF 2010)	85	85	85
Access to improved sanitation, %, 2008 (WHO and UNICEF 2010)	29	29	29
IBNET sourced data			
Number of utilities reporting in IBNET sample	1	1	1
Population served, water (thousands)	259	300	394
Size of the sample: total population living in service area, water supply (thousands)	550	560	571
Services coverage			
1.1 Water coverage (%)	47	54	69
2.1 Sewerage coverage (%)	9	13	16
Operational efficiency			
13.2 Electrical energy costs vs. operating costs (%) (share of energy cost as % of operational expenses)	—	—	—
6.1 Nonrevenue water (%)	28	30	28
6.2 Nonrevenue water (m³/km/day)	39.20	42.60	39.50
12.3 Staff W/1,000 W population served (W/1,000 W population served)	—	—	—
15.1 Continuity of service (hours/day) (duration of water supply, hours)	24.00	24.00	24.00
Financial efficiency			
8.1 Water sold that is metered (%)	—	—	—
23.1 Collection period (days)	210	226	234
23.2 Collection ratio (%)	—	—	—
18.1 Average revenue W & WW (US$/m³ water sold)	0.80	0.96	0.88
11.1 Operational cost W & WW (US$/m³ water sold)	0.71	0.79	0.85
24.1 Operating cost coverage (ratio)	1.13	1.21	1.04
Production and consumption			
3.1 Water production (l/person/day)	145	160	155
4.1 Total water consumption (l/person/day)	116	99	77
4.7 Residential consumption (l/person/day)	—	—	—
Poverty and affordability			
19.1 Total revenues/service population GNI (% GNI per capita) (average revenues)	4	4	3
19.2 Annual bill for households consuming 6m³ of water/month (US$/yr)	—	—	—
21.1 Ratio of industrial to residential tariff (level of cross-subsidy)	—	—	—

IBNET Indicator/Country: Liberia

Latest year available	2004	2005	2006
Surface area (km^2)	111,369	111,369	111,369
GNI per capita, Atlas method (current US$)	120	160	180
Total population	—	3,334,222	3,471,020
Urban population (%)	57	58	59
Total urban population	—	1,937,183	2,040,266
MDGs			
Access to improved water sources, %, 2008 (WHO and UNICEF 2010)	68	68	68
Access to improved sanitation, %, 2008 (WHO and UNICEF 2010)	17	17	17
IBNET sourced data			
Number of utilities reporting in IBNET sample	1	1	1
Population served, water (thousands)	350	350	350
Size of the sample: total population living in service area, water supply (thousands)	1,500	1,500	1,200
Services coverage			
1.1 Water coverage (%)	23	23	29
2.1 Sewerage coverage (%)	10	10	17
Operational efficiency			
13.2 Electrical energy costs vs. operating costs (%) (share of energy cost as % of operational expenses)	—	—	—
6.1 Nonrevenue water (%)	27	29	49
6.2 Nonrevenue water (m^3/km/day)	0.80	4.00	9.70
12.3 Staff W/1,000 W population served (W/1,000 W population served)	0.10	0.20	0.20
15.1 Continuity of service (hours/day) (duration of water supply, hours)	6.00	6.00	12.00
Financial efficiency			
8.1 Water sold that is metered (%)	—	—	95
23.1 Collection period (days)	80	133	127
23.2 Collection ratio (%)	57	63	75
18.1 Average revenue W & WW (US$/m^3 water sold)	1.15	1.15	1.22
11.1 Operational cost W & WW (US$/m^3 water sold)	0.91	1.17	1.17
24.1 Operating cost coverage (ratio)	1.26	0.98	1.05
Production and consumption			
3.1 Water production (l/person/day)	50	51	51
4.1 Total water consumption (l/person/day)	37	26	26
4.7 Residential consumption (l/person/day)	1	1	1
Poverty and affordability			
19.1 Total revenues/service population GNI (% GNI per capita) (average revenues)	2	2	1
19.2 Annual bill for households consuming 6m^3 of water/month (US$/yr)	44.59	45.33	48.00
21.1 Ratio of industrial to residential tariff (level of cross-subsidy)	4.15	3.40	2.54

IBNET Indicator/Country: Former Yugoslav Republic of Macedonia

Latest year available	2005	2006	2007
Surface area (km²)	25,713	25,713	25,713
GNI per capita, Atlas method (current US$)	2,500	2,600	2,700
Total population	2,035,312	2,037,863	2,039,838
Urban population (%)	65	66	66
Total urban population	1,331,094	1,342,952	1,354,452
MDGs			
Access to improved water sources, %, 2008 (WHO and UNICEF 2010)	100	100	100
Access to improved sanitation, %, 2008 (WHO and UNICEF 2010)	89	89	89
IBNET sourced data			
Number of utilities reporting in IBNET sample	15	15	15
Population served, water (thousands)	1,153	1,160	1,164
Size of the sample: total population living in service area, water supply (thousands)	1,209	1,219	1,222
Services coverage			
1.1 Water coverage (%)	95	95	95
2.1 Sewerage coverage (%)	78	79	79
Operational efficiency			
13.2 Electrical energy costs vs. operating costs (%) (share of energy cost as % of operational expenses)	11	10	11
6.1 Nonrevenue water (%)	59	60	60
6.2 Nonrevenue water (m³/km/day)	108.40	110.00	105.80
12.3 Staff W/1,000 W population served (W/1,000 W population served)	1.10	1.10	1.10
15.1 Continuity of service (hours/day) (duration of water supply, hours)	22.40	22.40	22.53
Financial efficiency			
8.1 Water sold that is metered (%)	94	94	94
23.1 Collection period (days)	479	512	486
23.2 Collection ratio (%)	87	85	83
18.1 Average revenue W & WW (US$/m³ water sold)	0.49	0.48	0.69
11.1 Operational cost W & WW (US$/m³ water sold)	0.29	0.32	0.36
24.1 Operating cost coverage (ratio)	1.69	1.50	1.91
Production and consumption			
3.1 Water production (l/person/day)	284	286	283
4.1 Total water consumption (l/person/day)	181	176	171
4.7 Residential consumption (l/person/day)	124	122	124
Poverty and affordability			
19.1 Total revenues/service population GNI (% GNI per capita) (average revenues)	1	1	2
19.2 Annual bill for households consuming 6m³ of water/month (US$/yr)	30.27	29.58	34.62
21.1 Ratio of industrial to residential tariff (level of cross-subsidy)	1.80	1.82	1.89

IBNET Indicator/Country: Madagascar

Latest year available	2003	2004	2005
Surface area (km^2)	587,041	587,041	587,041
GNI per capita, Atlas method (current US$)	280	290	330
Total population	—	—	17,614,261
Urban population (%)	28	28	29
Total urban population	—	—	5,020,064
MDGs			
Access to improved water sources, %, 2008 (WHO and UNICEF 2010)	41	41	41
Access to improved sanitation, %, 2008 (WHO and UNICEF 2010)	11	11	11
IBNET sourced data			
Number of utilities reporting in IBNET sample	1	1	1
Population served, water (thousands)	843	895	932
Size of the sample: total population living in service area, water supply (thousands)	961	986	1,032
Services coverage			
1.1 Water coverage (%)	88	91	90
2.1 Sewerage coverage (%)	20	20	20
Operational efficiency			
13.2 Electrical energy costs vs. operating costs (%) (share of energy cost as % of operational expenses)	—	—	—
6.1 Nonrevenue water (%)	36	33	34
6.2 Nonrevenue water (m^3/km/day)	33.50	29.40	30.10
12.3 Staff W/1,000 W population served (W/1,000 W population served)	—	—	—
15.1 Continuity of service (hours/day) (duration of water supply, hours)	—	—	—
Financial efficiency			
8.1 Water sold that is metered (%)	—	—	—
23.1 Collection period (days)	—	—	—
23.2 Collection ratio (%)	—	—	—
18.1 Average revenue W & WW (US$/m^3 water sold)	—	—	—
11.1 Operational cost W & WW (US$/m^3 water sold)	—	—	—
24.1 Operating cost coverage (ratio)	—	—	—
Production and consumption			
3.1 Water production (l/person/day)	294	293	296
4.1 Total water consumption (l/person/day)	195	192	187
4.7 Residential consumption (l/person/day)	—	—	—
Poverty and affordability			
19.1 Total revenues/service population GNI (% GNI per capita) (average revenues)	—	—	—
19.2 Annual bill for households consuming 6m^3 of water/month (US$/yr)	—	—	—
21.1 Ratio of industrial to residential tariff (level of cross-subsidy)	—	—	—

IBNET Indicator/Country: Malawi

Latest year available	2002	2003	2004
Surface area (km^2)	118,484	118,484	118,484
GNI per capita, Atlas method (current US$)	—	—	—
Total population	—	—	—
Urban population (%)	16	16	17
Total urban population	—	—	—
MDGs			
Access to improved water sources, %, 2008 (WHO and UNICEF 2010)	80	80	80
Access to improved sanitation, %, 2008 (WHO and UNICEF 2010)	56	56	56
IBNET sourced data			
Number of utilities reporting in IBNET sample	1	1	1
Population served, water (thousands)	891	951	1,014
Size of the sample: total population living in service area, water supply (thousands)	1,273	1,358	1,449
Services coverage			
1.1 Water coverage (%)	70	70	70
2.1 Sewerage coverage (%)	—	—	—
Operational efficiency			
13.2 Electrical energy costs vs. operating costs (%) (share of energy cost as % of operational expenses)	19	16	15
6.1 Nonrevenue water (%)	15	12	24
6.2 Nonrevenue water (m^3/km/day)	18.50	12.70	24.70
12.3 Staff W/1,000 W population served (W/1,000 W population served)	0.50	0.50	0.40
15.1 Continuity of service (hours/day) (duration of water supply, hours)	24.00	24.00	24.00
Financial efficiency			
8.1 Water sold that is metered (%)	100	100	100
23.1 Collection period (days)	98	128	132
23.2 Collection ratio (%)	90	92	91
18.1 Average revenue W & WW (US$/m^3 water sold)	0.54	0.35	0.26
11.1 Operational cost W & WW (US$/m^3 water sold)	0.37	0.18	0.16
24.1 Operating cost coverage (ratio)	1.46	2.00	1.70
Production and consumption			
3.1 Water production (l/person/day)	84	85	80
4.1 Total water consumption (l/person/day)	62	63	62
4.7 Residential consumption (l/person/day)	57	58	57
Poverty and affordability			
19.1 Total revenues/service population GNI (% GNI per capita) (average revenues)	8	6	4
19.2 Annual bill for households consuming 6m^3 of water/month (US$/yr)	60.44	40.51	30.19
21.1 Ratio of industrial to residential tariff (level of cross-subsidy)	8.48	8.46	8.44

IBNET Indicator/Country: Malaysia

Latest year available	2007
Surface area (km^2)	330,803
GNI per capita, Atlas method (current US$)	5,400
Total population	26,555,654
Urban population (%)	69
Total urban population	18,440,246
MDGs	
Access to improved water sources, %, 2008 (WHO and UNICEF 2010)	100
Access to improved sanitation, %, 2008 (WHO and UNICEF 2010)	96
IBNET sourced data	
Number of utilities reporting in IBNET sample	8
Population served, water (thousands)	17,442
Size of the sample: total population living in service area, water supply (thousands)	18,408
Services coverage	
1.1 Water coverage (%)	95
2.1 Sewerage coverage (%)	—
Operational efficiency	
13.2 Electrical energy costs vs. operating costs (%) (share of energy cost as % of operational expenses)	6
6.1 Nonrevenue water (%)	34
6.2 Nonrevenue water (m^3/km/day)	41.60
12.3 Staff W/1,000 W population served (W/1,000 W population served)	0.50
15.1 Continuity of service (hours/day) (duration of water supply, hours)	24.00
Financial efficiency	
8.1 Water sold that is metered (%)	100
23.1 Collection period (days)	365
23.2 Collection ratio (%)	—
18.1 Average revenue W & WW (US$/m^3 water sold)	0.39
11.1 Operational cost W & WW (US$/m^3 water sold)	0.34
24.1 Operating cost coverage (ratio)	1.15
Production and consumption	
3.1 Water production (l/person/day)	410
4.1 Total water consumption (l/person/day)	344
4.7 Residential consumption (l/person/day)	226
Poverty and affordability	
19.1 Total revenues/service population GNI (% GNI per capita) (average revenues)	1
19.2 Annual bill for households consuming 6m^3 of water/month (US$/yr)	8.25
21.1 Ratio of industrial to residential tariff (level of cross-subsidy)	1.73

IBNET Indicator/Country: Mali

Latest year available	2004	2005	2006
Surface area (km^2)	1,240,192	1,240,192	1,240,192
GNI per capita, Atlas method (current US$)	330	350	380
Total population	—	11,832,846	12,118,105
Urban population (%)	30	31	31
Total urban population	—	3,609,018	3,763,883
MDGs			
Access to improved water sources, %, 2008 (WHO and UNICEF 2010)	56	56	56
Access to improved sanitation, %, 2008 (WHO and UNICEF 2010)	36	36	36
IBNET sourced data			
Number of utilities reporting in IBNET sample	1	1	1
Population served, water (thousands)	1,497	1,653	1,682
Size of the sample: total population living in service area, water supply (thousands)	2,134	2,034	2,019
Services coverage			
1.1 Water coverage (%)	70	81	83
2.1 Sewerage coverage (%)	—	—	—
Operational efficiency			
13.2 Electrical energy costs vs. operating costs (%) (share of energy cost as % of operational expenses)	—	—	—
6.1 Nonrevenue water (%)	30	27	25
6.2 Nonrevenue water (m^3/km/day)	19.80	17.80	17.20
12.3 Staff W/1,000 W population served (W/1,000 W population served)	0.30	0.30	0.30
15.1 Continuity of service (hours/day) (duration of water supply, hours)	24.00	24.00	24.00
Financial efficiency			
8.1 Water sold that is metered (%)	100	100	100
23.1 Collection period (days)	—	—	—
23.2 Collection ratio (%)	—	—	—
18.1 Average revenue W & WW (US$/m^3 water sold)	0.51	0.56	0.60
11.1 Operational cost W & WW (US$/m^3 water sold)	0.32	0.30	0.32
24.1 Operating cost coverage (ratio)	1.62	1.87	1.88
Production and consumption			
3.1 Water production (l/person/day)	104	110	107
4.1 Total water consumption (l/person/day)	80	78	82
4.7 Residential consumption (l/person/day)	70	66	73
Poverty and affordability			
19.1 Total revenues/service population GNI (% GNI per capita) (average revenues)	5	5	5
19.2 Annual bill for households consuming 6m^3 of water/month (US$/yr)	—	—	—
21.1 Ratio of industrial to residential tariff (level of cross-subsidy)	1.26	2.65	4.24

IBNET Indicator/Country: Mauritania

Latest year available	2006	2007	2008
Surface area (km²)	1,025,520	1,025,520	1,025,520
GNI per capita, Atlas method (current US$)	550	570	590
Total population	3,062,283	3,138,922	3,215,043
Urban population (%)	41	41	41
Total urban population	1,243,287	1,280,680	1,318,168
MDGs			
Access to improved water sources, %, 2008 (WHO and UNICEF 2010)	49	49	49
Access to improved sanitation, %, 2008 (WHO and UNICEF 2010)	26	26	26
IBNET sourced data			
Number of utilities reporting in IBNET sample	1	1	1
Population served, water (thousands)	652	551	415
Size of the sample: total population living in service area, water supply (thousands)	973	1,404	1,476
Services coverage			
1.1 Water coverage (%)	67	39	28
2.1 Sewerage coverage (%)	—	—	—
Operational efficiency			
13.2 Electrical energy costs vs. operating costs (%) (share of energy cost as % of operational expenses)	—	22	—
6.1 Nonrevenue water (%)	36	34	38
6.2 Nonrevenue water (m³/km/day)	16.20	17.50	21.90
12.3 Staff W/1,000 W population served (W/1,000 W population served)	1.50	2.00	2.90
15.1 Continuity of service (hours/day) (duration of water supply, hours)	6.00	—	—
Financial efficiency			
8.1 Water sold that is metered (%)	100	100	100
23.1 Collection period (days)	321	765	—
23.2 Collection ratio (%)	116	84	—
18.1 Average revenue W & WW (US$/m³ water sold)	0.66	0.32	0.36
11.1 Operational cost W & WW (US$/m³ water sold)	0.72	0.71	—
24.1 Operating cost coverage (ratio)	0.91	0.45	—
Production and consumption			
3.1 Water production (l/person/day)	100	128	161
4.1 Total water consumption (l/person/day)	67	92	125
4.7 Residential consumption (l/person/day)	34	40	53
Poverty and affordability			
19.1 Total revenues/service population GNI (% GNI per capita) (average revenues)	3	2	3
19.2 Annual bill for households consuming 6m³ of water/month (US$/yr)	52.20	27.06	28.51
21.1 Ratio of industrial to residential tariff (level of cross-subsidy)	1.01	0.56	0.70

IBNET Indicator/Country: Mauritius

Latest year available	2004	2005	2006
Surface area (km^2)	2,040	2,040	2,040
GNI per capita, Atlas method (current US$)	4,640	5,000	5,100
Total population	—	1,243,253	1,252,987
Urban population (%)	42	42	42
Total urban population	—	525,896	530,765
MDGs			
Access to improved water sources, %, 2008 (WHO and UNICEF 2010)	99	99	99
Access to improved sanitation, %, 2008 (WHO and UNICEF 2010)	91	91	91
IBNET sourced data			
Number of utilities reporting in IBNET sample	1	1	1
Population served, water (thousands)	1,159	1,170	1,182
Size of the sample: total population living in service area, water supply (thousands)	1,159	1,171	1,182
Services coverage			
1.1 Water coverage (%)	100	100	100
2.1 Sewerage coverage (%)	—	—	—
Operational efficiency			
13.2 Electrical energy costs vs. operating costs (%) (share of energy cost as % of operational expenses)	—	—	—
6.1 Nonrevenue water (%)	53	52	54
6.2 Nonrevenue water (m^3/km/day)	57.60	56.20	62.10
12.3 Staff W/1,000 W population served (W/1,000 W population served)	0.90	0.80	0.80
15.1 Continuity of service (hours/day) (duration of water supply, hours)	24.00	24.00	24.00
Financial efficiency			
8.1 Water sold that is metered (%)	—	—	—
23.1 Collection period (days)	—	—	—
23.2 Collection ratio (%)	101	102	102
18.1 Average revenue W & WW (US$/m^3 water sold)	0.34	0.33	0.32
11.1 Operational cost W & WW (US$/m^3 water sold)	0.14	0.13	0.13
24.1 Operating cost coverage (ratio)	2.52	2.44	2.48
Production and consumption			
3.1 Water production (l/person/day)	414	437	440
4.1 Total water consumption (l/person/day)	205	212	212
4.7 Residential consumption (l/person/day)	160	164	164
Poverty and affordability			
19.1 Total revenues/service population GNI (% GNI per capita) (average revenues)	1	1	0
19.2 Annual bill for households consuming 6m^3 of water/month (US$/yr)	—	—	—
21.1 Ratio of industrial to residential tariff (level of cross-subsidy)	2.58	2.43	2.51

IBNET Indicator/Country: Mexico

Latest year available	2005	2006
Surface area (km²)	1,964,375	1,964,375
GNI per capita, Atlas method (current US$)	7,000	7,000
Total population	103,089,133	104,221,361
Urban population (%)	76	77
Total urban population	78,657,008	79,833,562
MDGs		
Access to improved water sources, %, 2008 (WHO and UNICEF 2010)	94	94
Access to improved sanitation, %, 2008 (WHO and UNICEF 2010)	85	85
IBNET sourced data		
Number of utilities reporting in IBNET sample	35	3
Population served, water (thousands)	12,941	669
Size of the sample: total population living in service area, water supply (thousands)	12,835	669
Services coverage		
1.1 Water coverage (%)	101	100
2.1 Sewerage coverage (%)	84	64
Operational efficiency		
13.2 Electrical energy costs vs. operating costs (%) (share of energy cost as % of operational expenses)	14	22
6.1 Nonrevenue water (%)	32	28
6.2 Nonrevenue water (m³/km/day)	35.30	22.70
12.3 Staff W/1,000 W population served (W/1,000 W population served)	—	—
15.1 Continuity of service (hours/day) (duration of water supply, hours)	21.11	23.10
Financial efficiency		
8.1 Water sold that is metered (%)	82	62
23.1 Collection period (days)	108	65
23.2 Collection ratio (%)	102	68
18.1 Average revenue W & WW (US$/m³ water sold)	0.74	0.73
11.1 Operational cost W & WW (US$/m³ water sold)	0.66	0.63
24.1 Operating cost coverage (ratio)	1.14	1.16
Production and consumption		
3.1 Water production (l/person/day)	240	248
4.1 Total water consumption (l/person/day)	164	179
4.7 Residential consumption (l/person/day)	121	141
Poverty and affordability		
19.1 Total revenues/service population GNI (% GNI per capita) (average revenues)	1	1
19.2 Annual bill for households consuming 6m³ of water/month (US$/yr)	—	52.07
21.1 Ratio of industrial to residential tariff (level of cross-subsidy)	1.99	—

IBNET Indicator/Country: Moldova

Latest year available	2006	2007	2008
Surface area (km^2)	33,846	33,846	33,846
GNI per capita, Atlas method (current US$)	750	770	1,000
Total population	3,708,848	3,667,469	3,633,369
Urban population (%)	42	42	42
Total urban population	1,569,584	1,541,804	1,517,295
MDGs			
Access to improved water sources, %, 2008 (WHO and UNICEF 2010)	—	—	—
Access to improved sanitation, %, 2008 (WHO and UNICEF 2010)	—	—	—
IBNET sourced data			
Number of utilities reporting in IBNET sample	41	41	39
Population served, water (thousands)	1,123	1,133	1,148
Size of the sample: total population living in service area, water supply (thousands)	1,377	1,409	1,446
Services coverage			
1.1 Water coverage (%)	82	80	80
2.1 Sewerage coverage (%)	65	63	64
Operational efficiency			
13.2 Electrical energy costs vs. operating costs (%) (share of energy cost as % of operational expenses)	26	22	23
6.1 Nonrevenue water (%)	42	43	42
6.2 Nonrevenue water (m^3/km/day)	32.80	36.30	33.40
12.3 Staff W/1,000 W population served (W/1,000 W population served)	2.80	2.60	2.60
15.1 Continuity of service (hours/day) (duration of water supply, hours)	16.03	17.39	17.78
Financial efficiency			
8.1 Water sold that is metered (%)	93	91	94
23.1 Collection period (days)	353	304	243
23.2 Collection ratio (%)	102	91	99
18.1 Average revenue W & WW (US$/m^3 water sold)	0.50	0.63	0.91
11.1 Operational cost W & WW (US$/m^3 water sold)	0.50	0.66	0.88
24.1 Operating cost coverage (ratio)	0.99	0.95	1.04
Production and consumption			
3.1 Water production (l/person/day)	216	225	245
4.1 Total water consumption (l/person/day)	148	158	156
4.7 Residential consumption (l/person/day)	110	116	114
Poverty and affordability			
19.1 Total revenues/service population GNI (% GNI per capita) (average revenues)	4	5	5
19.2 Annual bill for households consuming 6m^3 of water/month (US$/yr)	53.66	62.53	88.97
21.1 Ratio of industrial to residential tariff (level of cross-subsidy)	6.05	4.52	4.24

IBNET Indicator/Country: Mozambique

Latest year available	2005	2006	2007
Surface area (km^2)	801,590	801,590	801,590
GNI per capita, Atlas method (current US$)	290	300	330
Total population	20,834,379	21,353,466	21,869,362
Urban population (%)	35	35	36
Total urban population	7,187,861	7,533,503	7,886,092
MDGs			
Access to improved water sources, %, 2008 (WHO and UNICEF 2010)	47	47	47
Access to improved sanitation, %, 2008 (WHO and UNICEF 2010)	17	17	17
IBNET sourced data			
Number of utilities reporting in IBNET sample	5	5	5
Population served, water (thousands)	956	957	1,201
Size of the sample: total population living in service area, water supply (thousands)	3,166	3,199	3,211
Services coverage			
1.1 Water coverage (%)	30	30	37
2.1 Sewerage coverage (%)	—	—	—
Operational efficiency			
13.2 Electrical energy costs vs. operating costs (%) (share of energy cost as % of operational expenses)	18	17	—
6.1 Nonrevenue water (%)	58	56	59
6.2 Nonrevenue water (m^3/km/day)	128.00	120.30	131.20
12.3 Staff W/1,000 W population served (W/1,000 W population served)	1.20	1.20	1.00
15.1 Continuity of service (hours/day) (duration of water supply, hours)	14.40	16.20	19.20
Financial efficiency			
8.1 Water sold that is metered (%)	56	47	51
23.1 Collection period (days)	296	298	334
23.2 Collection ratio (%)	80	73	85
18.1 Average revenue W & WW (US$/m^3 water sold)	0.55	0.57	0.69
11.1 Operational cost W & WW (US$/m^3 water sold)	0.77	0.67	0.85
24.1 Operating cost coverage (ratio)	0.72	0.85	0.82
Production and consumption			
3.1 Water production (l/person/day)	225	221	200
4.1 Total water consumption (l/person/day)	103	106	87
4.7 Residential consumption (l/person/day)	—	—	—
Poverty and affordability			
19.1 Total revenues/service population GNI (% GNI per capita) (average revenues)	7	7	7
19.2 Annual bill for households consuming 6m^3 of water/month (US$/yr)	—	—	—
21.1 Ratio of industrial to residential tariff (level of cross-subsidy)	—	—	—

IBNET Indicator/Country: Namibia

Latest year available	2003	2004	2005
Surface area (km^2)	824,116	824,116	824,116
GNI per capita, Atlas method (current US$)	1,990	2,380	2,500
Total population	—	—	2,009,029
Urban population (%)	34	35	35
Total urban population	—	—	705,169
MDGs			
Access to improved water sources, %, 2008 (WHO and UNICEF 2010)	92	92	92
Access to improved sanitation, %, 2008 (WHO and UNICEF 2010)	33	33	33
IBNET sourced data			
Number of utilities reporting in IBNET sample	3	3	3
Population served, water (thousands)	287	296	304
Size of the sample: total population living in service area, water supply (thousands)	356	370	385
Services coverage			
1.1 Water coverage (%)	81	80	79
2.1 Sewerage coverage (%)	—	—	—
Operational efficiency			
13.2 Electrical energy costs vs. operating costs (%) (share of energy cost as % of operational expenses)	—	—	—
6.1 Nonrevenue water (%)	19	11	15
6.2 Nonrevenue water (m^3/km/day)	8.70	4.90	6.20
12.3 Staff W/1,000 W population served (W/1,000 W population served)	—	—	—
15.1 Continuity of service (hours/day) (duration of water supply, hours)	24.00	24.00	24.00
Financial efficiency			
8.1 Water sold that is metered (%)	100	100	100
23.1 Collection period (days)	649	641	627
23.2 Collection ratio (%)	—	—	—
18.1 Average revenue W & WW (US$/m^3 water sold)	1.38	1.63	1.85
11.1 Operational cost W & WW (US$/m^3 water sold)	1.09	1.45	2.05
24.1 Operating cost coverage (ratio)	1.24	1.07	0.88
Production and consumption			
3.1 Water production (l/person/day)	247	236	236
4.1 Total water consumption (l/person/day)	205	211	200
4.7 Residential consumption (l/person/day)	—	—	—
Poverty and affordability			
19.1 Total revenues/service population GNI (% GNI per capita) (average revenues)	—	—	—
19.2 Annual bill for households consuming 6m^3 of water/month (US$/yr)	—	—	—
21.1 Ratio of industrial to residential tariff (level of cross-subsidy)	—	—	—

IBNET Indicator/Country: Netherlands Antilles

Latest year available	2005	2006
Surface area (km^2)	37,354	37,354
GNI per capita, Atlas method (current US$)	33,000	33,000
Total population	186,451	189,102
Urban population (%)	92	92
Total urban population	171,348	174,276
MDGs		
Access to improved water sources, %, 2008 (WHO and UNICEF 2010)	—	—
Access to improved sanitation, %, 2008 (WHO and UNICEF 2010)	—	—
IBNET sourced data		
Number of utilities reporting in IBNET sample	1	1
Population served, water (thousands)	133	134
Size of the sample: total population living in service area, water supply (thousands)	133	134
Services coverage		
1.1 Water coverage (%)	100	100
2.1 Sewerage coverage (%)	—	—
Operational efficiency		
13.2 Electrical energy costs vs. operating costs (%) (share of energy cost as % of operational expenses)	—	—
6.1 Nonrevenue water (%)	30	29
6.2 Nonrevenue water (m^3/km/day)	4.20	4.00
12.3 Staff W/1,000 W population served (W/1,000 W population served)	2.70	2.60
15.1 Continuity of service (hours/day) (duration of water supply, hours)	24.00	24.00
Financial efficiency		
8.1 Water sold that is metered (%)	100	100
23.1 Collection period (days)	126	100
23.2 Collection ratio (%)	—	—
18.1 Average revenue W & WW (US$/m^3 water sold)	7.60	8.10
11.1 Operational cost W & WW (US$/m^3 water sold)	5.09	6.44
24.1 Operating cost coverage (ratio)	1.49	1.26
Production and consumption		
3.1 Water production (l/person/day)	240	242
4.1 Total water consumption (l/person/day)	179	179
4.7 Residential consumption (l/person/day)	132	128
Poverty and affordability		
19.1 Total revenues/service population GNI (% GNI per capita) (average revenues)	0	0
19.2 Annual bill for households consuming 6m^3 of water/month (US$/yr)	69.72	72.74
21.1 Ratio of industrial to residential tariff (level of cross-subsidy)	1.06	1.02

IBNET Indicator/Country: New Zealand

Latest year available	2005	2006	2007
Surface area (km^2)	270,467	270,467	270,467
GNI per capita, Atlas method (current US$)	21,000	22,000	24,000
Total population	4,133,900	4,184,600	4,228,300
Urban population (%)	86	86	86
Total urban population	3,563,422	3,612,147	3,654,943
MDGs			
Access to improved water sources, %, 2008 (WHO and UNICEF 2010)	100	100	100
Access to improved sanitation, %, 2008 (WHO and UNICEF 2010)	—	—	—
IBNET sourced data			
Number of utilities reporting in IBNET sample	1	1	1
Population served, water (thousands)	419	425	431
Size of the sample: total population living in service area, water supply (thousands)	419	425	431
Services coverage			
1.1 Water coverage (%)	100	100	100
2.1 Sewerage coverage (%)	100	102	100
Operational efficiency			
13.2 Electrical energy costs vs. operating costs (%) (share of energy cost as % of operational expenses)	—	—	—
6.1 Nonrevenue water (%)	18	10	12
6.2 Nonrevenue water (m^3/km/day)	12.40	1.00	1.40
12.3 Staff W/1,000 W population served (W/1,000 W population served)	—	—	—
15.1 Continuity of service (hours/day) (duration of water supply, hours)	24.00	24.00	24.00
Financial efficiency			
8.1 Water sold that is metered (%)	100	100	100
23.1 Collection period (days)	—	—	—
23.2 Collection ratio (%)	—	—	—
18.1 Average revenue W & WW (US$/m^3 water sold)	1.89	1.36	1.90
11.1 Operational cost W & WW (US$/m^3 water sold)	1.89	1.36	1.90
24.1 Operating cost coverage (ratio)	1.00	1.00	1.00
Production and consumption			
3.1 Water production (l/person/day)	355	385	390
4.1 Total water consumption (l/person/day)	294	349	342
4.7 Residential consumption (l/person/day)	189	164	168
Poverty and affordability			
19.1 Total revenues/service population GNI (% GNI per capita) (average revenues)	—	—	—
19.2 Annual bill for households consuming 6m^3 of water/month (US$/yr)	—	—	—
21.1 Ratio of industrial to residential tariff (level of cross-subsidy)	—	—	—

IBNET Indicator/Country: Nicaragua

Latest year available	2003	2004	2005
Surface area (km²)	130,373	130,373	130,373
GNI per capita, Atlas method (current US$)	770	830	850
Total population	—	—	5,455,216
Urban population (%)	55	56	56
Total urban population	—	—	3,049,466
MDGs			
Access to improved water sources, %, 2008 (WHO and UNICEF 2010)	85	85	85
Access to improved sanitation, %, 2008 (WHO and UNICEF 2010)	52	52	52
IBNET sourced data			
Number of utilities reporting in IBNET sample	1	1	1
Population served, water (thousands)	2,916	2,998	2,969
Size of the sample: total population living in service area, water supply (thousands)	3,190	3,153	3,153
Services coverage			
1.1 Water coverage (%)	91	95	94
2.1 Sewerage coverage (%)	35	35	34
Operational efficiency			
13.2 Electrical energy costs vs. operating costs (%) (share of energy cost as % of operational expenses)	—	—	40
6.1 Nonrevenue water (%)	57	—	40
6.2 Nonrevenue water (m³/km/day)	88.70	—	0.00
12.3 Staff W/1,000 W population served (W/1,000 W population served)	—	—	—
15.1 Continuity of service (hours/day) (duration of water supply, hours)	20.00	20.00	20.00
Financial efficiency			
8.1 Water sold that is metered (%)	—	—	69
23.1 Collection period (days)	—	—	151
23.2 Collection ratio (%)	—	—	82
18.1 Average revenue W & WW (US$/m³ water sold)	1.41	—	0.42
11.1 Operational cost W & WW (US$/m³ water sold)	—	—	0.38
24.1 Operating cost coverage (ratio)	—	—	1.11
Production and consumption			
3.1 Water production (l/person/day)	255	0	140
4.1 Total water consumption (l/person/day)	103	—	109
4.7 Residential consumption (l/person/day)	—	—	60
Poverty and affordability			
19.1 Total revenues/service population GNI (% GNI per capita) (average revenues)	—	—	2
19.2 Annual bill for households consuming 6m³ of water/month (US$/yr)	—	—	—
21.1 Ratio of industrial to residential tariff (level of cross-subsidy)	—	—	—

IBNET Indicator/Country: Niger

Latest year available	2003	2004	2005
Surface area (km²)	1,267,000	1,267,000	1,267,000
GNI per capita, Atlas method (current US$)	180	210	220
Total population	—	—	13,101,935
Urban population (%)	16	16	16
Total urban population	—	—	2,135,615
MDGs			
Access to improved water sources, %, 2008 (WHO and UNICEF 2010)	48	48	48
Access to improved sanitation, %, 2008 (WHO and UNICEF 2010)	9	9	9
IBNET sourced data			
Number of utilities reporting in IBNET sample	1	1	1
Population served, water (thousands)	1,253	1,502	1,613
Size of the sample: total population living in service area, water supply (thousands)	2,050	2,143	2,241
Services coverage			
1.1 Water coverage (%)	61	70	72
2.1 Sewerage coverage (%)	—	—	—
Operational efficiency			
13.2 Electrical energy costs vs. operating costs (%) (share of energy cost as % of operational expenses)			
6.1 Nonrevenue water (%)	17	17	19
6.2 Nonrevenue water (m³/km/day)	8.20	7.70	8.80
12.3 Staff W/1,000 W population served (W/1,000 W population served)	—	—	—
15.1 Continuity of service (hours/day) (duration of water supply, hours)	24.00	24.00	24.00
Financial efficiency			
8.1 Water sold that is metered (%)	100	100	100
23.1 Collection period (days)	43	250	193
23.2 Collection ratio (%)	—	—	—
18.1 Average revenue W & WW (US$/m³ water sold)	0.38	0.45	0.54
11.1 Operational cost W & WW (US$/m³ water sold)	0.39	0.26	0.42
24.1 Operating cost coverage (ratio)	0.98	1.70	1.30
Production and consumption			
3.1 Water production (l/person/day)	80	79	72
4.1 Total water consumption (l/person/day)	66	59	57
4.7 Residential consumption (l/person/day)	—	—	—
Poverty and affordability			
19.1 Total revenues/service population GNI (% GNI per capita) (average revenues)	5	5	5
19.2 Annual bill for households consuming 6m³ of water/month (US$/yr)	—	—	—
21.1 Ratio of industrial to residential tariff (level of cross-subsidy)	—	—	—

IBNET Indicator/Country: Nigeria

Latest year available	2002	2003	2004
Surface area (km^2)	923,768	923,768	923,768
GNI per capita, Atlas method (current US$)	320	380	430
Total population	—	—	—
Urban population (%)	44	45	45
Total urban population	—	—	—
MDGs			
Access to improved water sources, %, 2008 (WHO and UNICEF 2010)	58	58	58
Access to improved sanitation, %, 2008 (WHO and UNICEF 2010)	32	32	32
IBNET sourced data			
Number of utilities reporting in IBNET sample	12	12	12
Population served, water (thousands)	16,674	17,638	4,232
Size of the sample: total population living in service area, water supply (thousands)	29,242	30,709	31,761
Services coverage			
1.1 Water coverage (%)	57	57	48
2.1 Sewerage coverage (%)	—	—	—
Operational efficiency			
13.2 Electrical energy costs vs. operating costs (%) (share of energy cost as % of operational expenses)	30	27	30
6.1 Nonrevenue water (%)	31	31	50
6.2 Nonrevenue water (m^3/km/day)	191.20	193.40	47.50
12.3 Staff W/1,000 W population served (W/1,000 W population served)	0.50	0.50	0.80
15.1 Continuity of service (hours/day) (duration of water supply, hours)	12.00	12.10	8.67
Financial efficiency			
8.1 Water sold that is metered (%)	5	5	—
23.1 Collection period (days)	655	705	—
23.2 Collection ratio (%)	—	—	—
18.1 Average revenue W & WW (US$/m^3 water sold)	0.03	0.03	0.20
11.1 Operational cost W & WW (US$/m^3 water sold)	0.11	0.12	0.14
24.1 Operating cost coverage (ratio)	1.09	0.95	1.42
Production and consumption			
3.1 Water production (l/person/day)	291	295	287
4.1 Total water consumption (l/person/day)	182	174	140
4.7 Residential consumption (l/person/day)	69	72	75
Poverty and affordability			
19.1 Total revenues/service population GNI (% GNI per capita) (average revenues)	1	1	1
19.2 Annual bill for households consuming 6m^3 of water/month (US$/yr)	—	—	—
21.1 Ratio of industrial to residential tariff (level of cross-subsidy)	59.00	55.00	60.00

IBNET Indicator/Country: Pakistan

Latest year available	2004	2005	2006
Surface area (km²)	803,940	803,940	803,940
GNI per capita, Atlas method (current US$)	600	660	700
Total population	—	155,772,000	159,144,934
Urban population (%)	35	35	35
Total urban population	—	54,364,428	56,209,991
MDGs			
Access to improved water sources, %, 2008 (WHO and UNICEF 2010)	90	90	90
Access to improved sanitation, %, 2008 (WHO and UNICEF 2010)	45	45	45
IBNET sourced data			
Number of utilities reporting in IBNET sample	3	4	5
Population served, water (thousands)	3,033	4,343	8,753
Size of the sample: total population living in service area, water supply (thousands)	5,172	9,346	14,990
Services coverage			
1.1 Water coverage (%)	59	46	58
2.1 Sewerage coverage (%)	56	55	67
Operational efficiency			
13.2 Electrical energy costs vs. operating costs (%) (share of energy cost as % of operational expenses)	—	—	—
6.1 Nonrevenue water (%)	40	31	40
6.2 Nonrevenue water (m³/km/day)	81.80	63.30	67.90
12.3 Staff W/1,000 W population served (W/1,000 W population served)	—	—	—
15.1 Continuity of service (hours/day) (duration of water supply, hours)	8.33	9.25	10.60
Financial efficiency			
8.1 Water sold that is metered (%)	3	3	3
23.1 Collection period (days)	721	842	366
23.2 Collection ratio (%)	—	—	—
18.1 Average revenue W & WW (US$/m³ water sold)	0.08	0.07	0.17
11.1 Operational cost W & WW (US$/m³ water sold)	0.08	0.08	0.27
24.1 Operating cost coverage (ratio)	0.92	0.85	0.62
Production and consumption			
3.1 Water production (l/person/day)	233	264	213
4.1 Total water consumption (l/person/day)	127	115	97
4.7 Residential consumption (l/person/day)	—	—	—
Poverty and affordability			
19.1 Total revenues/service population GNI (% GNI per capita) (average revenues)	—	—	—
19.2 Annual bill for households consuming 6m³ of water/month (US$/yr)	—	—	—
21.1 Ratio of industrial to residential tariff (level of cross-subsidy)	—	—	—

IBNET Indicator/Country: Panama

Latest year available	2004	2005	2006
Surface area (km²)	75,517	75,517	75,517
GNI per capita, Atlas method (current US$)	4,210	4,300	4,400
Total population	—	3,231,624	3,287,575
Urban population (%)	70	71	72
Total urban population	—	2,287,990	2,353,904
MDGs			
Access to improved water sources, %, 2008 (WHO and UNICEF 2010)	93	93	93
Access to improved sanitation, %, 2008 (WHO and UNICEF 2010)	69	69	69
IBNET sourced data			
Number of utilities reporting in IBNET sample	1	1	1
Population served, water (thousands)	2,037	1,889	2,381
Size of the sample: total population living in service area, water supply (thousands)	2,243	2,303	2,372
Services coverage			
1.1 Water coverage (%)	91	82	100
2.1 Sewerage coverage (%)	52	45	48
Operational efficiency			
13.2 Electrical energy costs vs. operating costs (%) (share of energy cost as % of operational expenses)	60	38	44
6.1 Nonrevenue water (%)	44	43	39
6.2 Nonrevenue water (m³/km/day)	120.00	111.80	100.00
12.3 Staff W/1,000 W population served (W/1,000 W population served)	—	0.80	0.60
15.1 Continuity of service (hours/day) (duration of water supply, hours)	20.00	22.00	—
Financial efficiency			
8.1 Water sold that is metered (%)	44	46	43
23.1 Collection period (days)	342	152	112
23.2 Collection ratio (%)	—	78	111
18.1 Average revenue W & WW (US$/m³ water sold)	0.27	0.26	0.25
11.1 Operational cost W & WW (US$/m³ water sold)	0.11	0.19	0.18
24.1 Operating cost coverage (ratio)	2.38	1.39	1.44
Production and consumption			
3.1 Water production (l/person/day)	463	475	479
4.1 Total water consumption (l/person/day)	347	390	335
4.7 Residential consumption (l/person/day)	258	151	112
Poverty and affordability			
19.1 Total revenues/service population GNI (% GNI per capita) (average revenues)	1	1	1
19.2 Annual bill for households consuming 6m³ of water/month (US$/yr)	—	—	—
21.1 Ratio of industrial to residential tariff (level of cross-subsidy)	—	0.35	—

IBNET Indicator/Country: Paraguay

Latest year available	2003	2004	2005
Surface area (km^2)	406,752	406,752	406,752
GNI per capita, Atlas method (current US$)	1,070	1,140	1,200
Total population	—	—	5,904,155
Urban population (%)	57	58	59
Total urban population	—	—	3,453,931
MDGs			
Access to improved water sources, %, 2008 (WHO and UNICEF 2010)	86	86	86
Access to improved sanitation, %, 2008 (WHO and UNICEF 2010)	70	70	70
IBNET sourced data			
Number of utilities reporting in IBNET sample	1	4	4
Population served, water (thousands)	903	770	717
Size of the sample: total population living in service area, water supply (thousands)	1,188	1,421	1,002
Services coverage			
1.1 Water coverage (%)	76	54	72
2.1 Sewerage coverage (%)	43	39	32
Operational efficiency			
13.2 Electrical energy costs vs. operating costs (%) (share of energy cost as % of operational expenses)	19	19	17
6.1 Nonrevenue water (%)	52	45	44
6.2 Nonrevenue water (m^3/km/day)	44.70	43.80	41.80
12.3 Staff W/1,000 W population served (W/1,000 W population served)	—	—	0.50
15.1 Continuity of service (hours/day) (duration of water supply, hours)	24.00	24.00	24.00
Financial efficiency			
8.1 Water sold that is metered (%)	—	90	91
23.1 Collection period (days)	54	173	170
23.2 Collection ratio (%)	—	—	—
18.1 Average revenue W & WW (US$/m^3 water sold)	0.86	0.37	0.36
11.1 Operational cost W & WW (US$/m^3 water sold)	0.16	0.17	0.17
24.1 Operating cost coverage (ratio)	5.34	2.23	2.15
Production and consumption			
3.1 Water production (l/person/day)	260	374	400
4.1 Total water consumption (l/person/day)	130	242	243
4.7 Residential consumption (l/person/day)	121	206	205
Poverty and affordability			
19.1 Total revenues/service population GNI (% GNI per capita) (average revenues)	—	3	3
19.2 Annual bill for households consuming 6m^3 of water/month (US$/yr)	—	—	—
21.1 Ratio of industrial to residential tariff (level of cross-subsidy)	—	—	0.10

IBNET Indicator/Country: Peru

Latest year available	2004	2005	2006
Surface area (km^2)	1,285,216	1,285,216	1,285,216
GNI per capita, Atlas method (current US$)	2,360	2,500	2,700
Total population	—	27,835,927	28,175,982
Urban population (%)	71	71	71
Total urban population	—	19,791,344	20,061,299
MDGs			
Access to improved water sources, %, 2008 (WHO and UNICEF 2010)	82	82	82
Access to improved sanitation, %, 2008 (WHO and UNICEF 2010)	68	68	68
IBNET sourced data			
Number of utilities reporting in IBNET sample	9	10	50
Population served, water (thousands)	11,452	11,687	14,036
Size of the sample: total population living in service area, water supply (thousands)	13,071	13,658	16,765
Services coverage			
1.1 Water coverage (%)	88	86	84
2.1 Sewerage coverage (%)	78	79	75
Operational efficiency			
13.2 Electrical energy costs vs. operating costs (%) (share of energy cost as % of operational expenses)	—	—	—
6.1 Nonrevenue water (%)	41	43	43
6.2 Nonrevenue water (m^3/km/day)	61.70	61.50	63.10
12.3 Staff W/1,000 W population served (W/1,000 W population served)	—	0.30	—
15.1 Continuity of service (hours/day) (duration of water supply, hours)	16.50	15.78	16.06
Financial efficiency			
8.1 Water sold that is metered (%)	67	61	60
23.1 Collection period (days)	139	161	87
23.2 Collection ratio (%)	—	—	—
18.1 Average revenue W & WW (US$/m^3 water sold)	0.40	0.42	0.45
11.1 Operational cost W & WW (US$/m^3 water sold)	0.45	0.34	0.34
24.1 Operating cost coverage (ratio)	0.89	1.23	1.34
Production and consumption			
3.1 Water production (l/person/day)	219	201	266
4.1 Total water consumption (l/person/day)	130	134	140
4.7 Residential consumption (l/person/day)	36	—	—
Poverty and affordability			
19.1 Total revenues/service population GNI (% GNI per capita) (average revenues)	—	—	—
19.2 Annual bill for households consuming 6m^3 of water/month (US$/yr)	—	—	—
21.1 Ratio of industrial to residential tariff (level of cross-subsidy)	—	—	—

IBNET Indicator/Country: Philippines

Latest year available	2003	2004
Surface area (km²)	300,000	300,000
GNI per capita, Atlas method (current US$)	1,070	1,170
Total population	—	—
Urban population (%)	61	62
Total urban population	—	—
MDGs		
Access to improved water sources, %, 2008 (WHO and UNICEF 2010)	91	91
Access to improved sanitation, %, 2008 (WHO and UNICEF 2010)	76	76
IBNET sourced data		
Number of utilities reporting in IBNET sample	20	46
Population served, water (thousands)	301	10,608
Size of the sample: total population living in service area, water supply (thousands)	1,123	20,505
Services coverage		
1.1 Water coverage (%)	27	52
2.1 Sewerage coverage (%)	—	—
Operational efficiency		
13.2 Electrical energy costs vs. operating costs (%) (share of energy cost as % of operational expenses)	16	12
6.1 Nonrevenue water (%)	30	55
6.2 Nonrevenue water (m³/km/day)	27.90	207.70
12.3 Staff W/1,000 W population served (W/1,000 W population served)	1.40	0.70
15.1 Continuity of service (hours/day) (duration of water supply, hours)	20.10	20.98
Financial efficiency		
8.1 Water sold that is metered (%)	100	100
23.1 Collection period (days)	42	51
23.2 Collection ratio (%)	101	97
18.1 Average revenue W & WW (US$/m³ water sold)	0.22	0.28
11.1 Operational cost W & WW (US$/m³ water sold)	0.19	0.21
24.1 Operating cost coverage (ratio)	1.10	1.31
Production and consumption		
3.1 Water production (l/person/day)	159	267
4.1 Total water consumption (l/person/day)	123	193
4.7 Residential consumption (l/person/day)	108	140
Poverty and affordability		
19.1 Total revenues/service population GNI (% GNI per capita) (average revenues)	1	2
19.2 Annual bill for households consuming 6m³ of water/month (US$/yr)	—	—
21.1 Ratio of industrial to residential tariff (level of cross-subsidy)	2.09	2.29

IBNET Indicator/Country: Poland

Latest year available	2005	2006	2007
Surface area (km^2)	312,685	312,685	312,685
GNI per capita, Atlas method (current US$)	6,400	6,600	7,000
Total population	38,165,450	38,141,267	38,120,560
Urban population (%)	62	61	61
Total urban population	23,471,752	23,433,994	23,398,400
MDGs			
Access to improved water sources, %, 2008 (WHO and UNICEF 2010)	100	100	100
Access to improved sanitation, %, 2008 (WHO and UNICEF 2010)	90	90	90
IBNET sourced data			
Number of utilities reporting in IBNET sample	36	36	36
Population served, water (thousands)	9,269	9,251	9,247
Size of the sample: total population living in service area, water supply (thousands)	9,846	9,820	9,814
Services coverage			
1.1 Water coverage (%)	94	94	94
2.1 Sewerage coverage (%)	87	87	87
Operational efficiency			
13.2 Electrical energy costs vs. operating costs (%) (share of energy cost as % of operational expenses)	7	8	7
6.1 Nonrevenue water (%)	18	18	18
6.2 Nonrevenue water (m^3/km/day)	11.60	11.40	11.00
12.3 Staff W/1,000 W population served (W/1,000 W population served)	0.70	0.70	0.70
15.1 Continuity of service (hours/day) (duration of water supply, hours)	24.00	24.00	24.00
Financial efficiency			
8.1 Water sold that is metered (%)	99	99	99
23.1 Collection period (days)	58	62	67
23.2 Collection ratio (%)	97	98	98
18.1 Average revenue W & WW (US$/m^3 water sold)	1.35	1.39	1.71
11.1 Operational cost W & WW (US$/m^3 water sold)	1.12	1.16	1.47
24.1 Operating cost coverage (ratio)	1.20	1.20	1.16
Production and consumption			
3.1 Water production (l/person/day)	200	201	203
4.1 Total water consumption (l/person/day)	160	160	156
4.7 Residential consumption (l/person/day)	117	115	113
Poverty and affordability			
19.1 Total revenues/service population GNI (% GNI per capita) (average revenues)	1	1	1
19.2 Annual bill for households consuming 6m^3 of water/month (US$/yr)	64.87	66.28	82.82
21.1 Ratio of industrial to residential tariff (level of cross-subsidy)	1.29	1.26	1.26

IBNET Indicator/Country: Romania

Latest year available	2005	2006	2007
Surface area (km^2)	238,391	238,391	238,391
GNI per capita, Atlas method (current US$)	3,000	3,100	3,150
Total population	21,634,350	21,587,666	21,546,873
Urban population (%)	54	54	54
Total urban population	11,617,646	11,631,434	11,648,240
MDGs			
Access to improved water sources, %, 2008 (WHO and UNICEF 2010)	—	—	—
Access to improved sanitation, %, 2008 (WHO and UNICEF 2010)	72	72	72
IBNET sourced data			
Number of utilities reporting in IBNET sample	24	24	24
Population served, water (thousands)	3,675	3,712	3,889
Size of the sample: total population living in service area, water supply (thousands)	4,218	4,278	4,486
Services coverage			
1.1 Water coverage (%)	87	87	87
2.1 Sewerage coverage (%)	74	74	74
Operational efficiency			
13.2 Electrical energy costs vs. operating costs (%) (share of energy cost as % of operational expenses)	18	16	15
6.1 Nonrevenue water (%)	44	45	45
6.2 Nonrevenue water (m^3/km/day)	66.60	66.70	59.90
12.3 Staff W/1,000 W population served (W/1,000 W population served)	2.10	2.00	2.00
15.1 Continuity of service (hours/day) (duration of water supply, hours)	23.96	23.96	23.96
Financial efficiency			
8.1 Water sold that is metered (%)	88	90	93
23.1 Collection period (days)	103	97	85
23.2 Collection ratio (%)	104	106	104
18.1 Average revenue W & WW (US$/m^3 water sold)	0.64	0.78	1.08
11.1 Operational cost W & WW (US$/m^3 water sold)	0.53	0.63	0.92
24.1 Operating cost coverage (ratio)	1.20	1.24	1.18
Production and consumption			
3.1 Water production (l/person/day)	313	308	304
4.1 Total water consumption (l/person/day)	215	206	194
4.7 Residential consumption (l/person/day)	121	116	113
Poverty and affordability			
19.1 Total revenues/service population GNI (% GNI per capita) (average revenues)	2	2	2
19.2 Annual bill for households consuming 6m^3 of water/month (US$/yr)	74.06	75.17	91.31
21.1 Ratio of industrial to residential tariff (level of cross-subsidy)	1.56	1.59	1.48

IBNET Indicator/Country: Russian Federation

Latest year available	2006	2007	2008
Surface area (km^2)	17,098,242	17,098,242	17,098,242
GNI per capita, Atlas method (current US$)	5,780	7,560	5,780
Total population	142,500,000	142,100,000	141,950,000
Urban population (%)	73	73	73
Total urban population	103,854,000	103,534,060	103,396,380
MDGs			
Access to improved water sources, %, 2008 (WHO and UNICEF 2010)	96	96	96
Access to improved sanitation, %, 2008 (WHO and UNICEF 2010)	87	87	87
IBNET sourced data			
Number of utilities reporting in IBNET sample	84	84	80
Population served, water (thousands)	49,104	49,262	51,937
Size of the sample: total population living in service area, water supply (thousands)	51,718	51,945	51,937
Services coverage			
1.1 Water coverage (%)	95	95	100
2.1 Sewerage coverage (%)	94	94	94
Operational efficiency			
13.2 Electrical energy costs vs. operating costs (%) (share of energy cost as % of operational expenses)	16	23	13
6.1 Nonrevenue water (%)	22	20	20
6.2 Nonrevenue water (m^3/km/day)	67.10	57.90	56.20
12.3 Staff W/1,000 W population served (W/1,000 W population served)	1.50	1.50	1.30
15.1 Continuity of service (hours/day) (duration of water supply, hours)	24.00	24.00	24.00
Financial efficiency			
8.1 Water sold that is metered (%)	—	—	—
23.1 Collection period (days)	99	84	84
23.2 Collection ratio (%)	101	90	85
18.1 Average revenue W & WW (US$/m^3 water sold)	0.47	0.57	0.77
11.1 Operational cost W & WW (US$/m^3 water sold)	0.47	0.36	0.64
24.1 Operating cost coverage (ratio)	1.00	1.57	1.18
Production and consumption			
3.1 Water production (l/person/day)	504	480	451
4.1 Total water consumption (l/person/day)	412	403	357
4.7 Residential consumption (l/person/day)	247	242	221
Poverty and affordability			
19.1 Total revenues/service population GNI (% GNI per capita) (average revenues)	1	1	2
19.2 Annual bill for households consuming 6m^3 of water/month (US$/yr)	25.07	28.35	36.02
21.1 Ratio of industrial to residential tariff (level of cross-subsidy)	0.92	0.92	1.03

IBNET Indicator/Country: Rwanda

Latest year available	2003	2004	2005
Surface area (km^2)	26,338	26,338	26,338
GNI per capita, Atlas method (current US$)	200	210	220
Total population	—	—	8,992,140
Urban population (%)	16	17	18
Total urban population	—	—	1,573,625
MDGs			
Access to improved water sources, %, 2008 (WHO and UNICEF 2010)	65	65	65
Access to improved sanitation, %, 2008 (WHO and UNICEF 2010)	54	54	54
IBNET sourced data			
Number of utilities reporting in IBNET sample	1	1	1
Population served, water (thousands)	2,085	2,232	2,394
Size of the sample: total population living in service area, water supply (thousands)	1,843	1,973	2,010
Services coverage			
1.1 Water coverage (%)	100	100	100
2.1 Sewerage coverage (%)	—	—	—
Operational efficiency			
13.2 Electrical energy costs vs. operating costs (%) (share of energy cost as % of operational expenses)	42	31	46
6.1 Nonrevenue water (%)	51	44	38
6.2 Nonrevenue water (m^3/km/day)	11.50	8.70	7.10
12.3 Staff W/1,000 W population served (W/1,000 W population served)	0.60	0.60	0.60
15.1 Continuity of service (hours/day) (duration of water supply, hours)	12.00	12.00	12.00
Financial efficiency			
8.1 Water sold that is metered (%)	100	100	100
23.1 Collection period (days)	—	50	438
23.2 Collection ratio (%)	100	100	144
18.1 Average revenue W & WW (US$/m^3 water sold)	0.62	0.57	0.42
11.1 Operational cost W & WW (US$/m^3 water sold)	0.19	0.34	0.51
24.1 Operating cost coverage (ratio)	3.36	1.65	0.82
Production and consumption			
3.1 Water production (l/person/day)	24	17	16
4.1 Total water consumption (l/person/day)	12	11	11
4.7 Residential consumption (l/person/day)	—	—	—
Poverty and affordability			
19.1 Total revenues/service population GNI (% GNI per capita) (average revenues)	—	1	—
19.2 Annual bill for households consuming 6m^3 of water/month (US$/yr)	33.71	31.54	32.60
21.1 Ratio of industrial to residential tariff (level of cross-subsidy)	—	—	—

IBNET Indicator/Country: Senegal

Latest year available	2004	2005	2006
Surface area (km^2)	196,722	196,722	196,722
GNI per capita, Atlas method (current US$)	630	650	670
Total population	—	11,281,296	11,582,863
Urban population (%)	41	42	42
Total urban population	—	4,693,019	4,848,586
MDGs			
Access to improved water sources, %, 2008 (WHO and UNICEF 2010)	69	69	69
Access to improved sanitation, %, 2008 (WHO and UNICEF 2010)	51	51	51
IBNET sourced data			
Number of utilities reporting in IBNET sample	1	1	1
Population served, water (thousands)	4,191	4,440	4,597
Size of the sample: total population living in service area, water supply (thousands)	4,408	4,518	4,631
Services coverage			
1.1 Water coverage (%)	95	98	99
2.1 Sewerage coverage (%)	—	—	—
Operational efficiency			
13.2 Electrical energy costs vs. operating costs (%) (share of energy cost as % of operational expenses)	18	19	24
6.1 Nonrevenue water (%)	20	20	19
6.2 Nonrevenue water (m^3/km/day)	9.00	9.30	9.10
12.3 Staff W/1,000 W population served (W/1,000 W population served)	0.30	0.30	0.20
15.1 Continuity of service (hours/day) (duration of water supply, hours)	20.00	23.00	24.00
Financial efficiency			
8.1 Water sold that is metered (%)	—	—	—
23.1 Collection period (days)	61	57	59
23.2 Collection ratio (%)	98	98	99
18.1 Average revenue W & WW (US$/m^3 water sold)	0.81	0.81	0.87
11.1 Operational cost W & WW (US$/m^3 water sold)	0.65	0.63	0.68
24.1 Operating cost coverage (ratio)	1.25	1.28	1.28
Production and consumption			
3.1 Water production (l/person/day)	73	73	72
4.1 Total water consumption (l/person/day)	62	62	62
4.7 Residential consumption (l/person/day)	41	43	46
Poverty and affordability			
19.1 Total revenues/service population GNI (% GNI per capita) (average revenues)	3	3	3
19.2 Annual bill for households consuming 6m^3 of water/month (US$/yr)	21.73	21.76	22.08
21.1 Ratio of industrial to residential tariff (level of cross-subsidy)	1.26	1.32	1.42

IBNET Indicator/Country: Seychelles

Latest year available	2004	2005	2006
Surface area (km^2)	455	455	455
GNI per capita, Atlas method (current US$)	8,190	9,000	9,000
Total population	—	82,900	84,600
Urban population (%)	53	53	53
Total urban population	—	43,854	45,159
MDGs			
Access to improved water sources, %, 2008 (WHO and UNICEF 2010)	—	—	—
Access to improved sanitation, %, 2008 (WHO and UNICEF 2010)	—	—	—
IBNET sourced data			
Number of utilities reporting in IBNET sample	1	1	1
Population served, water (thousands)	79	80	80
Size of the sample: total population living in service area, water supply (thousands)	79	80	80
Services coverage			
1.1 Water coverage (%)	100	99	100
2.1 Sewerage coverage (%)	15	15	20
Operational efficiency			
13.2 Electrical energy costs vs. operating costs (%) (share of energy cost as % of operational expenses)	—	—	—
6.1 Nonrevenue water (%)	17	20	14
6.2 Nonrevenue water (m^3/km/day)	17.70	24.70	15.40
12.3 Staff W/1,000 W population served (W/1,000 W population served)	5.30	5.20	5.20
15.1 Continuity of service (hours/day) (duration of water supply, hours)	24.00	24.00	24.00
Financial efficiency			
8.1 Water sold that is metered (%)	50	45	45
23.1 Collection period (days)	—	—	—
23.2 Collection ratio (%)	99	100	100
18.1 Average revenue W & WW (US$/m^3 water sold)	1.06	0.77	0.79
11.1 Operational cost W & WW (US$/m^3 water sold)	2.04	1.69	1.77
24.1 Operating cost coverage (ratio)	0.52	0.45	0.44
Production and consumption			
3.1 Water production (l/person/day)	412	414	433
4.1 Total water consumption (l/person/day)	347	379	377
4.7 Residential consumption (l/person/day)	—	—	—
Poverty and affordability			
19.1 Total revenues/service population GNI (% GNI per capita) (average revenues)	2	1	1
19.2 Annual bill for households consuming 6m^3 of water/month (US$/yr)	86.60	81.96	83.00
21.1 Ratio of industrial to residential tariff (level of cross-subsidy)	—	—	—

IBNET Indicator/Country: Singapore

Latest year available	2007	2008
Surface area (km^2)	705	705
GNI per capita, Atlas method (current US$)	30,000	31,000
Total population	4,588,600	4,839,400
Urban population (%)	100	100
Total urban population	4,401,400	4,588,600
MDGs		
Access to improved water sources, %, 2008 (WHO and UNICEF 2010)	100	100
Access to improved sanitation, %, 2008 (WHO and UNICEF 2010)	100	100
IBNET sourced data		
Number of utilities reporting in IBNET sample	1	1
Population served, water (thousands)	4,589	4,840
Size of the sample: total population living in service area, water supply (thousands)	4,589	4,840
Services coverage		
1.1 Water coverage (%)	100	100
2.1 Sewerage coverage (%)	100	100
Operational efficiency		
13.2 Electrical energy costs vs. operating costs (%) (share of energy cost as % of operational expenses)	—	—
6.1 Nonrevenue water (%)	4	4
6.2 Nonrevenue water (m^3/km/day)	10.10	9.30
12.3 Staff W/1,000 W population served (W/1,000 W population served)	0.30	0.30
15.1 Continuity of service (hours/day) (duration of water supply, hours)	24.00	24.00
Financial efficiency		
8.1 Water sold that is metered (%)	100	100
23.1 Collection period (days)	—	—
23.2 Collection ratio (%)	—	—
18.1 Average revenue W & WW (US$/m^3 water sold)	—	—
11.1 Operational cost W & WW (US$/m^3 water sold)	—	—
24.1 Operating cost coverage (ratio)	—	—
Production and consumption		
3.1 Water production (l/person/day)	290	280
4.1 Total water consumption (l/person/day)	272	262
4.7 Residential consumption (l/person/day)	158	154
Poverty and affordability		
19.1 Total revenues/service population GNI (% GNI per capita) (average revenues)	—	—
19.2 Annual bill for households consuming 6m^3 of water/month (US$/yr)	90.54	101.74
21.1 Ratio of industrial to residential tariff (level of cross-subsidy)	—	—

IBNET Indicator/Country: Slovak Republic

Latest year available	2005	2006	2007
Surface area (km^2)	49,035	49,035	49,035
GNI per capita, Atlas method (current US$)	7,950	9,870	11,730
Total population	5,387,000	5,391,409	5,397,318
Urban population (%)	56	56	56
Total urban population	3,027,494	3,036,442	3,046,246
MDGs			
Access to improved water sources, %, 2008 (WHO and UNICEF 2010)	100	100	100
Access to improved sanitation, %, 2008 (WHO and UNICEF 2010)	100	100	100
IBNET sourced data			
Number of utilities reporting in IBNET sample	5	6	7
Population served, water (thousands)	2,732	3,430	3,664
Size of the sample: total population living in service area, water supply (thousands)	3,479	4,209	4,533
Services coverage			
1.1 Water coverage (%)	79	81	81
2.1 Sewerage coverage (%)	53	57	56
Operational efficiency			
13.2 Electrical energy costs vs. operating costs (%) (share of energy cost as % of operational expenses)	—	—	—
6.1 Nonrevenue water (%)	32	32	31
6.2 Nonrevenue water (m^3/km/day)	14.20	15.50	14.40
12.3 Staff W/1,000 W population served (W/1,000 W population served)	—	0.50	0.50
15.1 Continuity of service (hours/day) (duration of water supply, hours)	24.00	24.00	24.00
Financial efficiency			
8.1 Water sold that is metered (%)	100	100	100
23.1 Collection period (days)	72	70	80
23.2 Collection ratio (%)	104	103	102
18.1 Average revenue W & WW (US$/m^3 water sold)	1.34	1.50	1.81
11.1 Operational cost W & WW (US$/m^3 water sold)	0.94	1.04	1.28
24.1 Operating cost coverage (ratio)	1.43	1.44	1.42
Production and consumption			
3.1 Water production (l/person/day)	217	231	242
4.1 Total water consumption (l/person/day)	144	151	149
4.7 Residential consumption (l/person/day)	98	101	102
Poverty and affordability			
19.1 Total revenues/service population GNI (% GNI per capita) (average revenues)	1	1	1
19.2 Annual bill for households consuming 6m^3 of water/month (US$/yr)	91.13	112.34	133.96
21.1 Ratio of industrial to residential tariff (level of cross-subsidy)	1.32	0.98	0.98

IBNET Indicator/Country: South Africa

Latest year available	2004	2005	2006
Surface area (km²)	1,221,037	1,221,037	1,221,037
GNI per capita, Atlas method (current US$)	3,630	4,000	4,863
Total population	—	46,892,428	47,391,025
Urban population (%)	59	59	60
Total urban population	—	27,807,210	28,330,355
MDGs			
Access to improved water sources, %, 2008 (WHO and UNICEF 2010)	91	91	91
Access to improved sanitation, %, 2008 (WHO and UNICEF 2010)	77	77	77
IBNET sourced data			
Number of utilities reporting in IBNET sample	15	15	15
Population served, water (thousands)	15,905	16,297	16,633
Size of the sample: total population living in service area, water supply (thousands)	15,346	15,715	16,076
Services coverage			
1.1 Water coverage (%)	100	100	100
2.1 Sewerage coverage (%)	74	75	74
Operational efficiency			
13.2 Electrical energy costs vs. operating costs (%) (share of energy cost as % of operational expenses)	1	2	—
6.1 Nonrevenue water (%)	35	30	28
6.2 Nonrevenue water (m³/km/day)	20.10	21.60	19.50
12.3 Staff W/1,000 W population served (W/1,000 W population served)	—	0.40	0.40
15.1 Continuity of service (hours/day) (duration of water supply, hours)	24.00	24.00	24.00
Financial efficiency			
8.1 Water sold that is metered (%)	100	100	100
23.1 Collection period (days)	187	197	177
23.2 Collection ratio (%)	74	74	79
18.1 Average revenue W & WW (US$/m³ water sold)	1.11	1.09	1.07
11.1 Operational cost W & WW (US$/m³ water sold)	1.17	1.22	1.19
24.1 Operating cost coverage (ratio)	0.95	0.89	0.89
Production and consumption			
3.1 Water production (l/person/day)	297	288	271
4.1 Total water consumption (l/person/day)	180	188	186
4.7 Residential consumption (l/person/day)	136	146	145
Poverty and affordability			
19.1 Total revenues/service population GNI (% GNI per capita) (average revenues)	—	—	1
19.2 Annual bill for households consuming 6m³ of water/month (US$/yr)	—	2.00	1.00
21.1 Ratio of industrial to residential tariff (level of cross-subsidy)	—	—	—

IBNET Indicator/Country: **Sri Lanka**

Latest year available	2005	2006
Surface area (km²)	65,610	65,610
GNI per capita, Atlas method (current US$)	1,000	1,050
Total population	19,668,000	19,886,000
Urban population (%)	15	15
Total urban population	2,969,868	3,002,786
MDGs		
Access to improved water sources, %, 2008 (WHO and UNICEF 2010)	90	90
Access to improved sanitation, %, 2008 (WHO and UNICEF 2010)	91	91
IBNET sourced data		
Number of utilities reporting in IBNET sample	1	1
Population served, water (thousands)	1,686	1,722
Size of the sample: total population living in service area, water supply (thousands)	4,113	4,200
Services coverage		
1.1 Water coverage (%)	41	41
2.1 Sewerage coverage (%)	4	4
Operational efficiency		
13.2 Electrical energy costs vs. operating costs (%) (share of energy cost as % of operational expenses)	24	25
6.1 Nonrevenue water (%)	34	34
6.2 Nonrevenue water (m³/km/day)	35.50	37.50
12.3 Staff W/1,000 W population served (W/1,000 W population served)	—	—
15.1 Continuity of service (hours/day) (duration of water supply, hours)	16.00	20.00
Financial efficiency		
8.1 Water sold that is metered (%)	—	—
23.1 Collection period (days)	—	—
23.2 Collection ratio (%)	102	98
18.1 Average revenue W & WW (US$/m³ water sold)	0.23	0.22
11.1 Operational cost W & WW (US$/m³ water sold)	0.21	0.22
24.1 Operating cost coverage (ratio)	1.11	1.00
Production and consumption		
3.1 Water production (l/person/day)	583	598
4.1 Total water consumption (l/person/day)	412	416
4.7 Residential consumption (l/person/day)	270	275
Poverty and affordability		
19.1 Total revenues/service population GNI (% GNI per capita) (average revenues)	—	—
19.2 Annual bill for households consuming 6m³ of water/month (US$/yr)	—	—
21.1 Ratio of industrial to residential tariff (level of cross-subsidy)	—	—

IBNET Indicator/Country: Sudan

Latest year available	2005
Surface area (km^2)	2,505,813
GNI per capita, Atlas method (current US$)	550
Total population	38,698,472
Urban population (%)	15
Total urban population	3,021,510
MDGs	
Access to improved water sources, %, 2008 (WHO and UNICEF 2010)	57
Access to improved sanitation, %, 2008 (WHO and UNICEF 2010)	34
IBNET sourced data	
Number of utilities reporting in IBNET sample	1
Population served, water (thousands)	3,767
Size of the sample: total population living in service area, water supply (thousands)	9,903
Services coverage	
1.1 Water coverage (%)	38
2.1 Sewerage coverage (%)	—
Operational efficiency	
13.2 Electrical energy costs vs. operating costs (%) (share of energy cost as % of operational expenses)	20
6.1 Nonrevenue water (%)	53
6.2 Nonrevenue water (m^3/km/day)	111.90
12.3 Staff W/1,000 W population served (W/1,000 W population served)	1.00
15.1 Continuity of service (hours/day) (duration of water supply, hours)	17.33
Financial efficiency	
8.1 Water sold that is metered (%)	—
23.1 Collection period (days)	364
23.2 Collection ratio (%)	14
18.1 Average revenue W & WW (US$/m^3 water sold)	0.25
11.1 Operational cost W & WW (US$/m^3 water sold)	0.38
24.1 Operating cost coverage (ratio)	0.68
Production and consumption	
3.1 Water production (l/person/day)	198
4.1 Total water consumption (l/person/day)	88
4.7 Residential consumption (l/person/day)	—
Poverty and affordability	
19.1 Total revenues/service population GNI (% GNI per capita) (average revenues)	1
19.2 Annual bill for households consuming 6m^3 of water/month (US$/yr)	6.70
21.1 Ratio of industrial to residential tariff (level of cross-subsidy)	—

IBNET Indicator/Country: Swaziland

Latest year available	2007	2008	2009
Surface area (km²)	17,364	17,364	17,364
GNI per capita, Atlas method (current US$)	2,350	2,400	2,580
Total population	1,018,049	1,050,000	1,185,000
Urban population (%)	25	25	25
Total urban population	254,512	262,500	296,250
MDGs			
Access to improved water sources, %, 2008 (WHO and UNICEF 2010)	87	87	87
Access to improved sanitation, %, 2008 (WHO and UNICEF 2010)	50	50	50
IBNET sourced data			
Number of utilities reporting in IBNET sample	1	1	1
Population served, water (thousands)	270	285	285
Size of the sample: total population living in service area, water supply (thousands)	300	300	300
Services coverage			
1.1 Water coverage (%)	90	95	95
2.1 Sewerage coverage (%)	32	38	38
Operational efficiency			
13.2 Electrical energy costs vs. operating costs (%) (share of energy cost as % of operational expenses)	8	10	14
6.1 Nonrevenue water (%)	39.00	37.00	40.00
6.2 Nonrevenue water (m³/km/day)	28	26	30
12.3 Staff W/1,000 W population served (W/1,000 W population served)	1.39	1.25	1.25
15.1 Continuity of service (hours/day) (duration of water supply, hours)	24.00	24.00	24.00
Financial efficiency			
8.1 Water sold that is metered (%)	100	100	100
23.1 Collection period (days)	86	62	65
23.2 Collection ratio (%)	96	99	97
18.1 Average revenue W & WW (US$/m³ water sold)	1.53	1.40	1.56
11.1 Operational cost W & WW (US$/m³ water sold)	1.66	1.33	1.48
24.1 Operating cost coverage (ratio)	0.92	1.05	1.05
Production and consumption			
3.1 Water production (l/person/day)	183.00	183.00	192.00
4.1 Total water consumption (l/person/day)	112.00	115.00	115.00
4.7 Residential consumption (l/person/day)	78	77	77
Poverty and affordability			
19.1 Total revenues/service population GNI (% GNI per capita) (average revenues)	2	2	3
19.2 Annual bill for households consuming 6m³ of water/month (US$/yr)	79.10	74.75	79.37
21.1 Ratio of industrial to residential tariff (level of cross-subsidy)	N/A	3.00	3.02

IBNET Indicator/Country: Tajikistan

Latest year available	2003	2004	2005
Surface area (km^2)	143,100	143,100	143,100
GNI per capita, Atlas method (current US$)	210	280	330
Total population	—	—	6,535,538
Urban population (%)	26	26	26
Total urban population	—	—	1,725,382
MDGs			
Access to improved water sources, %, 2008 (WHO and UNICEF 2010)	70	70	70
Access to improved sanitation, %, 2008 (WHO and UNICEF 2010)	94	94	94
IBNET sourced data			
Number of utilities reporting in IBNET sample	9	9	9
Population served, water (thousands)	1,029	1,042	1,090
Size of the sample: total population living in service area, water supply (thousands)	1,112	1,128	1,179
Services coverage			
1.1 Water coverage (%)	93	92	92
2.1 Sewerage coverage (%)	60	60	59
Operational efficiency			
13.2 Electrical energy costs vs. operating costs (%) (share of energy cost as % of operational expenses)	—	—	—
6.1 Nonrevenue water (%)	35	35	36
6.2 Nonrevenue water (m^3/km/day)	198.60	207.60	225.90
12.3 Staff W/1,000 W population served (W/1,000 W population served)	1.50	1.40	1.20
15.1 Continuity of service (hours/day) (duration of water supply, hours)	22.40	21.73	21.29
Financial efficiency			
8.1 Water sold that is metered (%)	1	2	1
23.1 Collection period (days)	326	263	273
23.2 Collection ratio (%)	52	47	42
18.1 Average revenue W & WW (US$/m^3 water sold)	0.01	0.02	0.03
11.1 Operational cost W & WW (US$/m^3 water sold)	0.01	0.01	0.02
24.1 Operating cost coverage (ratio)	1.39	1.40	1.42
Production and consumption			
3.1 Water production (l/person/day)	700	718	744
4.1 Total water consumption (l/person/day)	518	542	554
4.7 Residential consumption (l/person/day)	313	328	336
Poverty and affordability			
19.1 Total revenues/service population GNI (% GNI per capita) (average revenues)	1	1	2
19.2 Annual bill for households consuming 6m^3 of water/month (US$/yr)	0.87	1.66	1.62
21.1 Ratio of industrial to residential tariff (level of cross-subsidy)	20.09	22.21	23.56

IBNET Indicator/Country: Tanzania

Latest year available	2006	2007	2008
Surface area (km^2)	945,087	945,087	945,087
GNI per capita, Atlas method (current US$)	360	400	410
Total population	40,117,243	41,276,209	42,483,923
Urban population (%)	25	25	26
Total urban population	9,884,889	10,352,073	10,841,897
MDGs			
Access to improved water sources, %, 2008 (WHO and UNICEF 2010)	—	—	—
Access to improved sanitation, %, 2008 (WHO and UNICEF 2010)	—	—	—
IBNET sourced data			
Number of utilities reporting in IBNET sample	19	20	20
Population served, water (thousands)	0	4,749	5,675
Size of the sample: total population living in service area, water supply (thousands)	0	5,977	6,959
Services coverage			
1.1 Water coverage (%)	—	79	82
2.1 Sewerage coverage (%)	—	5	4
Operational efficiency			
13.2 Electrical energy costs vs. operating costs (%) (share of energy cost as % of operational expenses)	—	25	—
6.1 Nonrevenue water (%)	41	45	36
6.2 Nonrevenue water (m^3/km/day)	32.10	52.00	34.80
12.3 Staff W/1,000 W population served (W/1,000 W population served)	—	0.50	0.50
15.1 Continuity of service (hours/day) (duration of water supply, hours)	18.11	15.00	17.58
Financial efficiency			
8.1 Water sold that is metered (%)	—	100	100
23.1 Collection period (days)	109	81	95
23.2 Collection ratio (%)	103	88	97
18.1 Average revenue W & WW (US$/m^3 water sold)	0.28	0.35	0.24
11.1 Operational cost W & WW (US$/m^3 water sold)	0.28	0.40	0.29
24.1 Operating cost coverage (ratio)	0.98	0.88	0.83
Production and consumption			
3.1 Water production (l/person/day)	100	121	128
4.1 Total water consumption (l/person/day)	—	64	61
4.7 Residential consumption (l/person/day)	—	43	—
Poverty and affordability			
19.1 Total revenues/service population GNI (% GNI per capita) (average revenues)	—	2	—
19.2 Annual bill for households consuming 6m^3 of water/month (US$/yr)	—	—	—
21.1 Ratio of industrial to residential tariff (level of cross-subsidy)	—	1.02	—

IBNET Indicator/Country: Togo

Latest year available	2002	2003	2004
Surface area (km^2)	56,785	56,785	56,785
GNI per capita, Atlas method (current US$)	240	260	310
Total population	5,500,000	5,775,000	6,063,750
Urban population (%)	35	35	35
Total urban population	1,925,000	2,021,250	2,122,313
MDGs			
Access to improved water sources, %, 2008 (WHO and UNICEF 2010)	87	87	87
Access to improved sanitation, %, 2008 (WHO and UNICEF 2010)	42	42	42
IBNET sourced data			
Number of utilities reporting in IBNET sample	1	1	1
Population served, water (thousands)	1,008	1,022	1,097
Size of the sample: total population living in service area, water supply (thousands)	2,169	2,249	2,332
Services coverage			
1.1 Water coverage (%)	46	45	47
2.1 Sewerage coverage (%)	6	5	7
Operational efficiency			
13.2 Electrical energy costs vs. operating costs (%) (share of energy cost as % of operational expenses)	11	16	12
6.1 Nonrevenue water (%)	27.00	24.00	28.00
6.2 Nonrevenue water (m^3/km/day)	8	7	8
12.3 Staff W/1,000 W population served (W/1,000 W population served)	0.68	0.64	0.57
15.1 Continuity of service (hours/day) (duration of water supply, hours)	24.00	24.00	24.00
Financial efficiency			
8.1 Water sold that is metered (%)	92	91	91
23.1 Collection period (days)	N/A	N/A	N/A
23.2 Collection ratio (%)	87	72	54
18.1 Average revenue W & WW (US$/m^3 water sold)	0.65	0.82	0.67
11.1 Operational cost W & WW (US$/m^3 water sold)	0.77	0.65	0.96
24.1 Operating cost coverage (ratio)	0.85	1.26	0.69
Production and consumption			
3.1 Water production (l/person/day)	56.16	55.26	51.39
4.1 Total water consumption (l/person/day)	41.00	42.00	37.00
4.7 Residential consumption (l/person/day)	24	24	24
Poverty and affordability			
19.1 Total revenues/service population GNI (% GNI per capita) (average revenues)	4	5	3
19.2 Annual bill for households consuming 6m^3 of water/month (US$/yr)	46.93	57.16	60.46
21.1 Ratio of industrial to residential tariff (level of cross-subsidy)	1.82	0.87	2.01

IBNET Indicator/Country: Tunisia

Latest year available	2004	2005	2006
Surface area (km²)	163,610	163,610	163,610
GNI per capita, Atlas method (current US$)	2,650	2,800	2,850
Total population	—	10,029,000	10,128,100
Urban population (%)	65	65	66
Total urban population	—	6,548,937	6,654,162
MDGs			
Access to improved water sources, %, 2008 (WHO and UNICEF 2010)	94	94	94
Access to improved sanitation, %, 2008 (WHO and UNICEF 2010)	85	85	85
IBNET sourced data			
Number of utilities reporting in IBNET sample	1	1	1
Population served, water (thousands)	10,000	10,100	10,200
Size of the sample: total population living in service area, water supply (thousands)	10,000	10,100	10,200
Services coverage			
1.1 Water coverage (%)	100	100	100
2.1 Sewerage coverage (%)	—	—	—
Operational efficiency			
13.2 Electrical energy costs vs. operating costs (%) (share of energy cost as % of operational expenses)	—	—	—
6.1 Nonrevenue water (%)	22	22	23
6.2 Nonrevenue water (m³/km/day)	6.10	6.30	6.60
12.3 Staff W/1,000 W population served (W/1,000 W population served)	—	—	—
15.1 Continuity of service (hours/day) (duration of water supply, hours)	24.00	24.00	24.00
Financial efficiency			
8.1 Water sold that is metered (%)	—	—	—
23.1 Collection period (days)	212	222	242
23.2 Collection ratio (%)	100	100	100
18.1 Average revenue W & WW (US$/m³ water sold)	0.42	0.41	0.40
11.1 Operational cost W & WW (US$/m³ water sold)	0.46	0.52	0.50
24.1 Operating cost coverage (ratio)	0.92	0.79	0.80
Production and consumption			
3.1 Water production (l/person/day)	103	120	139
4.1 Total water consumption (l/person/day)	86	88	91
4.7 Residential consumption (l/person/day)	59	61	64
Poverty and affordability			
19.1 Total revenues/service population GNI (% GNI per capita) (average revenues)	1	0	0
19.2 Annual bill for households consuming 6m³ of water/month (US$/yr)	—	—	—
21.1 Ratio of industrial to residential tariff (level of cross-subsidy)	2.26	2.13	2.13

IBNET Indicator/Country: Turkey

Latest year available	2006	2007	2008
Surface area (km^2)	783,562	783,562	783,562
GNI per capita, Atlas method (current US$)	4,500	4,750	5,000
Total population	72,087,928	73,003,736	73,914,260
Urban population (%)	68	68	69
Total urban population	48,846,780	49,803,149	50,764,314
MDGs			
Access to improved water sources, %, 2008 (WHO and UNICEF 2010)	99	99	99
Access to improved sanitation, %, 2008 (WHO and UNICEF 2010)	90	90	90
IBNET sourced data			
Number of utilities reporting in IBNET sample	20	20	20
Population served, water (thousands)	2,398	2,619	905
Size of the sample: total population living in service area, water supply (thousands)	2,328	2,641	990
Services coverage			
1.1 Water coverage (%)	100	99	100
2.1 Sewerage coverage (%)	94	95	94
Operational efficiency			
13.2 Electrical energy costs vs. operating costs (%) (share of energy cost as % of operational expenses)	54	51	41
6.1 Nonrevenue water (%)	56	62	59
6.2 Nonrevenue water (m^3/km/day)	43.80	62.60	43.40
12.3 Staff W/1,000 W population served (W/1,000 W population served)	—	0.70	0.50
15.1 Continuity of service (hours/day) (duration of water supply, hours)	24.00	24.00	24.00
Financial efficiency			
8.1 Water sold that is metered (%)	97	96	100
23.1 Collection period (days)	—	139	108
23.2 Collection ratio (%)	96	94	90
18.1 Average revenue W & WW (US$/m^3 water sold)	1.04	1.36	1.21
11.1 Operational cost W & WW (US$/m^3 water sold)	0.72	1.09	0.93
24.1 Operating cost coverage (ratio)	1.37	1.25	1.25
Production and consumption			
3.1 Water production (l/person/day)	181	233	192
4.1 Total water consumption (l/person/day)	97	92	108
4.7 Residential consumption (l/person/day)	72	73	87
Poverty and affordability			
19.1 Total revenues/service population GNI (% GNI per capita) (average revenues)	1	1	1
19.2 Annual bill for households consuming 6m^3 of water/month (US$/yr)	66.05	79.77	85.00
21.1 Ratio of industrial to residential tariff (level of cross-subsidy)	2.31	2.25	1.72

IBNET Indicator/Country: Uganda

Latest year available	2007	2008	2009
Surface area (km^2)	241,038	241,038	241,038
GNI per capita, Atlas method (current US$)	300	310	350
Total population	30,637,544	31,656,865	—
Urban population (%)	13	13	—/
Total urban population	3,927,733	4,109,061	—
MDGs			
Access to improved water sources, %, 2008 (WHO and UNICEF 2010)	67	67	67
Access to improved sanitation, %, 2008 (WHO and UNICEF 2010)	48	48	48
IBNET sourced data			
Number of utilities reporting in IBNET sample	1	1	1
Population served, water (thousands)	1,803	1,944	2,137
Size of the sample: total population living in service area, water supply (thousands)	2,540	2,700	2,946
Services coverage			
1.1 Water coverage (%)	71	72	73
2.1 Sewerage coverage (%)	7	6	6
Operational efficiency			
13.2 Electrical energy costs vs. operating costs (%) (share of energy cost as % of operational expenses)	—	18	18
6.1 Nonrevenue water (%)	33	34	36
6.2 Nonrevenue water (m^3/km/day)	16.90	17.50	14.40
12.3 Staff W/1,000 W population served (W/1,000 W population served)	0.70	0.60	0.60
15.1 Continuity of service (hours/day) (duration of water supply, hours)	23.00	23.00	23.00
Financial efficiency			
8.1 Water sold that is metered (%)	—	—	—
23.1 Collection period (days)	518	525	449
23.2 Collection ratio (%)	92	92	99
18.1 Average revenue W & WW (US$/m^3 water sold)	1.05	1.29	1.10
11.1 Operational cost W & WW (US$/m^3 water sold)	0.78	1.04	0.82
24.1 Operating cost coverage (ratio)	1.34	1.24	1.34
Production and consumption			
3.1 Water production (l/person/day)	101	104	109
4.1 Total water consumption (l/person/day)	62	60	57
4.7 Residential consumption (l/person/day)	33	31	30
Poverty and affordability			
19.1 Total revenues/service population GNI (% GNI per capita) (average revenues)	8	9	7
19.2 Annual bill for households consuming 6m^3 of water/month (US$/yr)	—	—	—
21.1 Ratio of industrial to residential tariff (level of cross-subsidy)	—	—	—

IBNET Indicator/Country: Ukraine

Latest year available	2005	2006	2007
Surface area (km^2)	603,500	603,500	603,500
GNI per capita, Atlas method (current US$)	1,500	1,750	1,900
Total population	47,105,150	46,787,750	46,509,350
Urban population (%)	68	68	68
Total urban population	31,937,292	31,750,167	31,589,151
MDGs			
Access to improved water sources, %, 2008 (WHO and UNICEF 2010)	98	98	98
Access to improved sanitation, %, 2008 (WHO and UNICEF 2010)	95	95	95
IBNET sourced data			
Number of utilities reporting in IBNET sample	16	16	16
Population served, water (thousands)	2,703	2,721	2,736
Size of the sample: total population living in service area, water supply (thousands)	3,452	3,432	3,411
Services coverage			
1.1 Water coverage (%)	78	79	80
2.1 Sewerage coverage (%)	63	64	67
Operational efficiency			
13.2 Electrical energy costs vs. operating costs (%) (share of energy cost as % of operational expenses)	30	33	36
6.1 Nonrevenue water (%)	42	43	44
6.2 Nonrevenue water (m^3/km/day)	76.50	76.80	75.10
12.3 Staff W/1,000 W population served (W/1,000 W population served)	2.20	2.10	2.10
15.1 Continuity of service (hours/day) (duration of water supply, hours)	22.00	22.00	22.00
Financial efficiency			
8.1 Water sold that is metered (%)	27	31	36
23.1 Collection period (days)	278	251	225
23.2 Collection ratio (%)	92	84	92
18.1 Average revenue W & WW (US$/m^3 water sold)	0.25	0.32	0.44
11.1 Operational cost W & WW (US$/m^3 water sold)	0.30	0.37	0.48
24.1 Operating cost coverage (ratio)	0.84	0.87	0.91
Production and consumption			
3.1 Water production (l/person/day)	530	520	506
4.1 Total water consumption (l/person/day)	311	296	283
4.7 Residential consumption (l/person/day)	231	224	208
Poverty and affordability			
19.1 Total revenues/service population GNI (% GNI per capita) (average revenues)	2	2	2
19.2 Annual bill for households consuming 6m^3 of water/month (US$/yr)	18.07	29.49	37.32
21.1 Ratio of industrial to residential tariff (level of cross-subsidy)	5.30	4.93	3.25

IBNET Indicator/Country: Uruguay

Latest year available	2003	2004	2006
Surface area (km²)	176,215	176,215	176,215
GNI per capita, Atlas method (current US$)	3,780	3,900	5,812
Total population	—	—	3,314,466
Urban population (%)	92	92	92
Total urban population	—	—	3,052,623
MDGs			
Access to improved water sources, %, 2008 (WHO and UNICEF 2010)	100	100	100
Access to improved sanitation, %, 2008 (WHO and UNICEF 2010)	100	100	100
IBNET sourced data			
Number of utilities reporting in IBNET sample	1	1	1
Population served, water (thousands)	3,064	2,834	3,055
Size of the sample: total population living in service area, water supply (thousands)	3,178	3,101	—
Services coverage			
1.1 Water coverage (%)	96	91	94
2.1 Sewerage coverage (%)	15	17	22
Operational efficiency			
13.2 Electrical energy costs vs. operating costs (%) (share of energy cost as % of operational expenses)	—	—	—
6.1 Nonrevenue water (%)	52	54	54
6.2 Nonrevenue water (m³/km/day)	34.90	37.10	35.70
12.3 Staff W/1,000 W population served (W/1,000 W population served)	—	—	—
15.1 Continuity of service (hours/day) (duration of water supply, hours)	24.00	24.00	24.00
Financial efficiency			
8.1 Water sold that is metered (%)	97	96	97
23.1 Collection period (days)	67	55	45
23.2 Collection ratio (%)	—	—	97
18.1 Average revenue W & WW (US$/m³ water sold)	0.98	1.04	1.36
11.1 Operational cost W & WW (US$/m³ water sold)	0.61	0.65	0.82
24.1 Operating cost coverage (ratio)	1.61	1.62	1.66
Production and consumption			
3.1 Water production (l/person/day)	275	288	320
4.1 Total water consumption (l/person/day)	117	128	133
4.7 Residential consumption (l/person/day)	92	100	104
Poverty and affordability			
19.1 Total revenues/service population GNI (% GNI per capita) (average revenues)	1	1	1
19.2 Annual bill for households consuming 6m³ of water/month (US$/yr)	—	—	—
21.1 Ratio of industrial to residential tariff (level of cross-subsidy)	2.47	2.39	—

IBNET Indicator/Country: Uzbekistan

Latest year available	2005	2006	2007
Surface area (km^2)	447,400	447,400	447,400
GNI per capita, Atlas method (current US$)	470	500	520
Total population	26,167,369	26,485,800	26,867,800
Urban population (%)	37	37	37
Total urban population	9,603,424	9,730,883	9,881,977
MDGs			
Access to improved water sources, %, 2008 (WHO and UNICEF 2010)	87	87	87
Access to improved sanitation, %, 2008 (WHO and UNICEF 2010)	100	100	100
IBNET sourced data			
Number of utilities reporting in IBNET sample	5	5	3
Population served, water (thousands)	3,101	3,091	2,423
Size of the sample: total population living in service area, water supply (thousands)	4,100	4,097	3,418
Services coverage			
1.1 Water coverage (%)	76	75	71
2.1 Sewerage coverage (%)	14	15	6
Operational efficiency			
13.2 Electrical energy costs vs. operating costs (%) (share of energy cost as % of operational expenses)	18	21	27
6.1 Nonrevenue water (%)	46	39	29
6.2 Nonrevenue water (m^3/km/day)	35.20	25.10	7.60
12.3 Staff W/1,000 W population served (W/1,000 W population served)	1.20	1.20	1.10
15.1 Continuity of service (hours/day) (duration of water supply, hours)	20.40	20.00	17.33
Financial efficiency			
8.1 Water sold that is metered (%)	4	6	8
23.1 Collection period (days)	286	260	255
23.2 Collection ratio (%)	95	102	109
18.1 Average revenue W & WW (US$/m^3 water sold)	0.06	0.07	0.11
11.1 Operational cost W & WW (US$/m^3 water sold)	0.09	0.10	0.15
24.1 Operating cost coverage (ratio)	0.75	0.69	0.75
Production and consumption			
3.1 Water production (l/person/day)	255	218	77
4.1 Total water consumption (l/person/day)	122	119	61
4.7 Residential consumption (l/person/day)	72	67	49
Poverty and affordability			
19.1 Total revenues/service population GNI (% GNI per capita) (average revenues)	1	1	0
19.2 Annual bill for households consuming 6m^3 of water/month (US$/yr)	13.61	13.82	13.97
21.1 Ratio of industrial to residential tariff (level of cross-subsidy)	4.72	6.40	5.44

IBNET Indicator/Country: República Bolivariana de Venezuela

Latest year available	2006
Surface area (km^2)	912,050
GNI per capita, Atlas method (current US$)	6,070
Total population	27,031,000
Urban population (%)	93
Total urban population	25,041,518
MDGs	
Access to improved water sources, %, 2008 (WHO and UNICEF 2010)	—
Access to improved sanitation, %, 2008 (WHO and UNICEF 2010)	—
IBNET sourced data	
Number of utilities reporting in IBNET sample	17
Population served, water (thousands)	22,677
Size of the sample: total population living in service area, water supply (thousands)	25,149
Services coverage	
1.1 Water coverage (%)	90
2.1 Sewerage coverage (%)	74
Operational efficiency	
13.2 Electrical energy costs vs. operating costs (%) (share of energy cost as % of operational expenses)	41
6.1 Nonrevenue water (%)	62
6.2 Nonrevenue water (m^3/km/day)	137.80
12.3 Staff W/1,000 W population served (W/1,000 W population served)	0.60
15.1 Continuity of service (hours/day) (duration of water supply, hours)	20.00
Financial efficiency	
8.1 Water sold that is metered (%)	38
23.1 Collection period (days)	416
23.2 Collection ratio (%)	91
18.1 Average revenue W & WW (US$/m^3 water sold)	0.25
11.1 Operational cost W & WW (US$/m^3 water sold)	0.26
24.1 Operating cost coverage (ratio)	0.95
Production and consumption	
3.1 Water production (l/person/day)	369
4.1 Total water consumption (l/person/day)	178
4.7 Residential consumption (l/person/day)	128
Poverty and affordability	
19.1 Total revenues/service population GNI (% GNI per capita) (average revenues)	0
19.2 Annual bill for households consuming 6m^3 of water/month (US$/yr)	—
21.1 Ratio of industrial to residential tariff (level of cross-subsidy)	—

IBNET Indicator/Country: Vietnam

Latest year available	2005	2006	2007
Surface area (km^2)	331,212	331,212	331,212
GNI per capita, Atlas method (current US$)	600	620	650
Total population	83,106,300	84,136,800	85,154,900
Urban population (%)	26	27	27
Total urban population	21,940,063	22,615,972	23,298,381
MDGs			
Access to improved water sources, %, 2008 (WHO and UNICEF 2010)	94	94	94
Access to improved sanitation, %, 2008 (WHO and UNICEF 2010)	75	75	75
IBNET sourced data			
Number of utilities reporting in IBNET sample	68	68	68
Population served, water (thousands)	14,871	16,326	17,806
Size of the sample: total population living in service area, water supply (thousands)	21,650	22,430	24,400
Services coverage			
1.1 Water coverage (%)	65	69	69
2.1 Sewerage coverage (%)	40	33	33
Operational efficiency			
13.2 Electrical energy costs vs. operating costs (%) (share of energy cost as % of operational expenses)	32	36	35
6.1 Nonrevenue water (%)	35	34	32
6.2 Nonrevenue water (m^3/km/day)	32.60	28.60	26.50
12.3 Staff W/1,000 W population served (W/1,000 W population served)	1.30	1.20	1.20
15.1 Continuity of service (hours/day) (duration of water supply, hours)	21.51	21.58	21.79
Financial efficiency			
8.1 Water sold that is metered (%)	100	100	100
23.1 Collection period (days)	355	367	329
23.2 Collection ratio (%)	99	99	99
18.1 Average revenue W & WW (US$/m^3 water sold)	0.23	0.24	0.24
11.1 Operational cost W & WW (US$/m^3 water sold)	0.12	0.12	0.13
24.1 Operating cost coverage (ratio)	1.87	1.92	1.88
Production and consumption			
3.1 Water production (l/person/day)	212	256	266
4.1 Total water consumption (l/person/day)	140	139	142
4.7 Residential consumption (l/person/day)	93	93	94
Poverty and affordability			
19.1 Total revenues/service population GNI (% GNI per capita) (average revenues)	2	2	2
19.2 Annual bill for households consuming 6m^3 of water/month (US$/yr)	1.95	2.33	2.45
21.1 Ratio of industrial to residential tariff (level of cross-subsidy)	3.50	3.48	3.76

IBNET Indicator/Country: Zambia

Latest year available	2007	2008	2009
Surface area (km^2)	752,612	752,612	752,612
GNI per capita, Atlas method (current US$)	460	500	510
Total population	12,019,481	12,313,942	12,620,219
Urban population (%)	35	35	—
Total urban population	4,344,359	4,470,082	—
MDGs			
Access to improved water sources, %, 2008 (WHO and UNICEF 2010)	60	60	60
Access to improved sanitation, %, 2008 (WHO and UNICEF 2010)	49	49	49
IBNET sourced data			
Number of utilities reporting in IBNET sample	10	10	10
Population served, water (thousands)	3,284	3,305	3,612
Size of the sample: total population living in service area, water supply (thousands)	4,640	4,691	4,811
Services coverage			
1.1 Water coverage (%)	71	70	75
2.1 Sewerage coverage (%)	34	29	33
Operational efficiency			
13.2 Electrical energy costs vs. operating costs (%) (share of energy cost as % of operational expenses)	—	—	—
6.1 Nonrevenue water (%)	46	45	45
6.2 Nonrevenue water (m^3/km/day)	—	—	—
12.3 Staff W/1,000 W population served (W/1,000 W population served)	—	—	—
15.1 Continuity of service (hours/day) (duration of water supply, hours)	15.00	15.50	16.20
Financial efficiency			
8.1 Water sold that is metered (%)	—	—	—
23.1 Collection period (days)	—	—	—
23.2 Collection ratio (%)	83	91	76
18.1 Average revenue W & WW (US$/m^3 water sold)	0.33	0.44	0.33
11.1 Operational cost W & WW (US$/m^3 water sold)	0.27	0.38	0.28
24.1 Operating cost coverage (ratio)	1.23	1.15	1.16
Production and consumption			
3.1 Water production (l/person/day)	316	318	326
4.1 Total water consumption (l/person/day)	141	144	135
4.7 Residential consumption (l/person/day)	—	—	—
Poverty and affordability			
19.1 Total revenues/service population GNI (% GNI per capita) (average revenues)	—	—	—
19.2 Annual bill for households consuming 6m^3 of water/month (US$/yr)	20.08	23.64	17.21
21.1 Ratio of industrial to residential tariff (level of cross-subsidy)	—	—	—

Appendix 3.   IBNET Indicators

Service Coverage

Indicators

Indicator	Unit	Concept
1.1 Water Coverage	%	Population with easy access to water services (either with direct service connection or within reach of a public water point)/total population under utility's nominal responsibility, expressed in percentage
1.2 *Water Coverage—Household Connections*	%	*Subset of 1.1*
1.3 *Water Coverage—Public Water Points*	%	*Subset of 1.1*
2.1 Sewerage Coverage	%	Population with sewerage services (direct service connection)/total population under utility's notional responsibility, expressed in percentage

Source: Authors.

Discussion

Coverage is a key development indicator. All coverage indicators are affected by whether the data on populations and household sizes is up to date and accurate. The need to estimate populations served by public-water points may affect the confidence that can be placed in the water coverage measure. In the *Global Water Supply and Sanitation Assessment 2000 Report* (WHO and UNICEF 2000), reasonable access was defined as "the availability of at least 20 liters per person per day from a source within one kilometer of the user's dwelling." However, we recommend that the population within 250 meters be used as a rule of thumb.

Water Consumption and Production

Indicators

Indicator	Unit	Concept
3.1 Water Production	liters/person/day	Total annual water supplied to the distribution system (including purchased water, if any) expressed by population served per day and by connection per month
3.2 Water Production	m^3/conn/month	
4.1 Total Water Consumption	liters/person/day	Total annual water sold, expressed by population served per day and by connection per month
4.2 Total Water Consumption	m^3/conn/month	
Water consumption split by customer type:	%	*Shows the split of total water consumption into four categories of customer*
4.3 Residential Consumption *4.4 Industrial/Commercial Consumption* *4.5 Consumption by Institutions and Others* *4.6 Bulk Treated Supply*		

Indicator	Unit	Concept
Residential Consumption:		
4.7 Residential Consumption	*liters/person/day*	*Shows the average water consumption of groups of people*
4.8 Residential Consumption —Connections to Main Supply		
4.9 Residential Consumption —Public Water Points		

Source: Authors.
Note: m³/conn/month = cubic meters per connection per month.

Discussion

Theoretically, the most accurate water consumption indicator would be expressed in terms of liters/person/day. This indicator presents data problems, however, notably lack of accurate total consumption data (for example, from universal metering) and poor quality or outdated census data.

While the accuracy of service population figures may need improvement, utilities are often more confident of the number of connections in their system. In addition, water production figures may be known more reliably than water consumption figures.

To draw on these other sources of (potentially) more reliable data, we have included a number of indicators that allow utilities to undertake trending analyses. Interutility comparisons are more difficult, however, given the different mix of household sizes and of multiple dwellings served by a single connection. This is especially the case between utilities in different countries. In-country comparisons will be more accurate due to the homogeneity of household size and of dwellings per connection.

Nonrevenue Water

Indicators

Indicator	Unit	Concept
6.1 Nonrevenue water	%	Difference between water supplied and water sold
6.2 Nonrevenue water	m³/km/day	expressed as a percentage of net water supplied;
6.3 Nonrevenue water	m³/conn/day	as volume of water "lost" per kilometer of water-distribution network per day; and volume of water "lost" per water connection per day

Source: Authors.
Note: m³/conn/day = cubic meter per connection per day, m³/km/day = cubic meters per kilometer per day.

Discussion

Nonrevenue water represents water that has been produced but is either "lost" before it reaches the customer (through leaks or theft) or used legally but not paid for. Some nonrevenue water can be recovered by appropriate technical or managerial actions and then used to meet currently unsatisfied demand (thus increasing revenues to the utility); such recovery helps defer future capital expenditures for the provision of additional supply, thus reducing costs to the utility.

The International Water Association (IWA) distinguishes between *nonrevenue water* (percent) and *unaccounted-for water,* which does not include legal but unpaid usage and which is usually measured in cubic meters per connection per day. The difference is usually small, and here, only the term *nonrevenue water* is used.

The most appropriate measure for quantifying unaccounted-for water is the subject of ongoing debate. A percentage approach can make utilities with high levels of consumption or compact networks appear to be better performing than those with low levels of consumption or extensive networks. For a capture of these different perspectives, reporting three measures of unaccounted-for water has become the norm.

Meters

Indicators

Indicator	Unit	Concept
7.1 Metering Level	%	Total number of connections with operating meter/total number of connections, expressed in percentage
8.1 Water Sold that Is Metered	%	Volume of water sold that is metered/total volume of water sold, expressed in percentage

Source: Authors.

Discussion

Metering customers' water use is considered good practice. It allows customers the opportunity to influence their water bills, and it provides utilities with tools and information that allow them to better manage their systems.

The indicators provide two separate perspectives on the issue, each of which is significant individually as well as in conjunction with the other. Together the indicators provide insight into the effectiveness of a metering installation strategy; the ratio of indicator (8)/(7) indicates the extent to which a utility is targeting large water users as its highest priority.

Network Performance

Indicators

Indicator	Unit	Concept
9.1 Pipe Breaks	breaks/km/yr	Total number of pipe breaks per year expressed per kilometer of the water-distribution network.
10.1 Sewer System Blockages	blockages/km/yr	Total number of blockages per year expressed per kilometer of sewers.

Source: Authors.

Discussion

The number of pipe breaks, relative to the scale of the system, indicates the ability of the pipe network to provide service to customers. The rate of water-pipe breaks can also be seen as a surrogate for the general state of the network, although it reflects operation and maintenance practices as well as physical condition. Highly aggregated reporting, however, can conceal that some sections of the network may fail repeatedly while much of the remainder is in reasonable condition. Break rates itemized by different materials, diameters, or time laid can show if and where breaks are concentrated.

Sewer blockages, likewise, measure the sewer network's ability to provide service to customers. Blockages can reflect a number of problems, including the effectiveness of routine operations and maintenance activities, the hydraulic performance of the network, and the general condition of the pipes.

As defined here, *water-pipe bursts* may occur at one of three places: on mains; in service pipes, which are the utility's responsibility; or at joints or fittings. They may be found through visible signs of water in addition to leak detection by utility staff. *Sewer blockages* include all blockages or collapses occurring in all sewers or drains for which the utility has responsibility, whatever action is needed to clear them.

Operating Costs and Staff

Indicators

Indicator	Unit	Concept
11.1 Unit Operational Cost W and WW	US$/m^3 sold	Total annual operational expenses[a]/total annual volume sold
11.2 Unit Operational Cost W and WW	US$/m^3 produced	Total annual operational expenses[a]/total annual water produced
11.3 Unit Operational Cost—Water Only	US$/m^3 sold	Annual water service operational expenses[a]/total annual volume sold
11.4 Operational Cost Split—% Water	%	Split of the total cost into water and wastewater
11.5 Operational Cost Split—% Wastewater	%	
11.6 Unit Operational Cost—Wastewater	US$/WW pop served	Annual wastewater operational expenses[a]/population served
12.2 Staff W and WW/1,000 W and WW conn	#/1,000 W and WW conn	Total number of staff expressed as per thousand connections
12.1 Staff W/1,000 W conn	#/1,000 W conn	
12.5 Staff WW/1,000 WW conn	#/1,000 WW conn	
12.4 Staff W & WW/1,000 W and WW Pop Served	#/1,000 W & WW pop served	Total number of staff expressed as per thousand people served
12.3 Staff W/1,000 W Pop Served	#/1,000 W pop served	
12.6 Staff WW/1,000 WW Pop Served	#/1,000 WW pop served	
12.7 Staff % Water	%	
12.8 Staff % Wastewater	%	
13.1 Labor Costs vs. Operational Costs	%	Total annual labor costs (including benefits) expressed as a percentage of total annual operational costs
13.2 Electrical Energy Costs vs. Operational Costs	%	Annual electrical energy costs expressed as a percentage of total annual operational costs
14.1 Contracted-Out service Costs versus Operational Costs	%	Total cost of services contracted out to the private sector expressed as a percentage of total annual operational[a] costs

Source: Authors.
Note: conn = connection, Pop = population, W = water, WW = wastewater, # = number, a: Annual operating expenses exclude depreciation, interest, and debt service.

Discussion

Unit operational costs provide a bottom-line assessment of the mix of resources used to achieve the outputs required. The preferred denominator related to operational costs is the amount of water sold. This ratio then reflects the cost of providing water at the customer take-off point.

Lack of universal metering, the doubtful accuracy of many household meters, and a focus in the past on water production indicates that an alternative measure of operational cost per cubic meter of water produced is also relevant in the short term.

Staff costs are traditionally a major component of operating costs. Understanding staffing levels can often provide a quick indicator of the extent of any overstaffing in a water utility. While allocating staff time to either water or wastewater services is useful, this information is sometimes not available. Comparisons are best made between utilities offering the same scope of service in terms both of total size and of mix of water and sewer services. Staff number comparisons should reflect any extensive use of outside contractors (see indicator 14.1).

The number of people served per connection varies from country to country and from utility to utility, depending on the housing stock and the approach taken toward service connection. For facilitation of international comparisons, a denominator of populations served has been included here.

The relative importance of staff costs compared to total costs is captured in indicator 13.1. Utilities are often overstaffed, and this measure shows the impact of possible changes in future staff numbers.

Electrical power costs are often important (indicator 13.2), as when, for example, power has been very cheap and used inefficiently.

Quality of Service

Indicators

Indicator	Unit	Concept
15.1 Continuity of Service	hours/day	Average hours of service per day for water supply
15.2 Customers with Discontinuous Supply	%	Percentage of customers with a water supply that is discontinuous during normal operation
15.3 Quality of Water Supplied: Number of Tests for Residual Chlorine	% of # required	Number of tests carried out on samples taken from the distribution system, as a percentage of the number required by the applicable standard, which may exceed 100 percent Operational samples, or any others not taken to check compliance with the standard, are excluded
15.4 Quality of Water Supplied: Samples Passing on Residual Chlorine	%	Percentage of samples tested for residual chlorine that pass the relevant standard
16.1 Complaints about W and WW Services	% of W and WW conn	Total number of water and wastewater complaints per year expressed as a percentage of the total number of water and wastewater connections
17.1 Wastewater—At Least Primary Treatment	%	Proportion of collected sewage that receives at least primary treatment, that is, involving settlement with the intention of removing solids but not biological treatment. Both lagoon and mechanical treatment can be included, where appropriate.
17.2 Wastewater—Primary Treatment Only	%	Proportion of collected sewage that receives primary treatment only, that is, involving settlement with the intention of removing solids but not biological treatment. Both lagoon and mechanical treatment can be included, where appropriate.
17.3 Wastewater—Secondary Treatment or Better	%	Proportion of collected sewage that receives at least secondary treatment, that is, removing oxygen demand as well as solids, usually using biological methods. Both lagoon and mechanical treatment can be included, where appropriate.

Source: Authors.
Note: W = water, WW = wastewater, # = number.

Discussion

Historically, limited attention has been paid to measures that capture the quality of service provided to customers. This, in fact, should be a particular focus of performance measurement.

The measures presented above are a limited first step in the process of capturing information on quality of service. Complaints, while relatively easy to track, give only a glimpse of actual company performance, since consumers may have become accustomed to poor service and no longer complain about it. In other instances, customers may find it difficult to report complaints. Capturing at least some customer-derived data, however, is considered an important starting point for evaluating quality.

Because wastewater is collected does not mean it is fully treated before its discharge back to the environment. The wastewater treatment indicators provide an understanding of the amount of effluent being treated before discharge.

A more comprehensive set of service-quality indicators could be developed, but in the short term it is unlikely that utility managers will collect the necessary data. Expansion of the indicator set is therefore a medium- to long-term objective.

Billings and Collections

Indicators

Indicator	Unit	Concept
18.1 Average Revenue W & WW	US$/m^3 water sold	Total annual water and wastewater operating revenues expressed by annual amount of water sold and by the number of connections.
18.2 Average Revenue W & WW	US$/W conn/yr	
18.3 Average Revenue—Water Only	US$/m^3 water sold	Operating revenues (W only) expressed by annual amount of water sold.
18.4 Revenue Split—% Water	% of total for W & WW	The percentage split of total revenue into water and wastewater.
18.5 Revenue Split—% Wastewater		
18.6 Water Revenue—Residential	% of total water revenue	The percentage split of water revenue by customer type.
18.7 Water Revenue—Industrial/Commercial		
18.8 Water Revenue—Institutions & Others		
18.9 Water Revenue—Bulk-Treated Supply		
18.10 WW Revenue	US$/person served	Operating revenues (wastewater only) expressed per person served.
19.1 Total Revenues per Service Pop/GNI	% GNI per capita	Total annual operating revenues per population served/national GNI per capita; expressed in percentage.
19.2 Monthly Water Bill (for a household consuming 6m^3 of water per month through a household or shared yard tap, but excluding the use of standpipe)	US$/yr	Cost in local currency to a household per month of 6m^3 water/exchange rate with US$ × 12
20.1 Residential Fixed Component of Tariff		Any fixed component of the residential tariff (total amount).
20.3 Residential Fixed Component of Tariff—Water	US$/conn/yr	Water and wastewater together, separated if possible.
20.4 Residential Fixed Component of Tariff—Wastewater		

Indicator	Unit	Concept
20.2 Residential Fixed Component of Tariff		Any fixed component of the residential tariff as a proportion of the average tariff per connection per year
20.5 Residential Fixed Component of Tariff—Water	% of average bill	Water and wastewater together, separated if possible
20.6 *Residential Fixed Component of Tariff—Wastewater*		
21.1 Ratio of Industrial to Residential Tariff	ratio	The average charge (per cubic meter) to industrial customers compared to the average charge (per cubic meter) to residential customers
21.2 *Ratio of Industrial to Residential Tariff—Water*		*Water and wastewater together, separated if possible*
21.3 *Ratio of Industrial to Residential Tariff—Wastewater*		
22.1 Connection Charge—Water	US$/conn	The cost to make a residential pipe connection to the water system and the sewer system, measured in absolute amount and as a proportion of national GNI per capita
22.2 Connection Charge—Water	% GNI per capita	
22.3 Connection Charge—Sewerage	US$/conn	
22.4 Connection Charge—Sewerage	% GNI per capita	
23.1 Collection Period	days	(Year-end accounts receivable/total annual operating revenues) × 365
23.2 Collection Ratio	%	Cash income/billed revenue as a %

Source: Authors.

Note: conn = connection, GNI = gross national income, m^3 = cubic meter, W = water service, WW = wastewater or sewerage service.

Discussion

Average tariffs must be put in the perspective of affordability. Household income data, however, is not easy to obtain. The indicator selected here, therefore, compares average per capita tariffs as a proportion of per capita gross national income (GNI). GNI represents the entire country, without reflecting local variations, but it is the most appropriate and consistent measure available for the majority of countries. Here, the GNI should be that calculated using the Atlas method.

Some utilities use fixed-charge components within the residential tariff (that is, not taking account of the amount of water consumed). Such tariffs can adversely affect low-volume water consumers, but they also protect the utility's revenue stream during periods of highly variable consumption. Comparison of the fixed component with the average tariff will indicate the relative weight of the fixed and variable components in a water bill.

Cross-subsidies may exist between industrial consumers and residential consumers. The ratio of the average charges (per cubic meter) to industrial and residential customers provides some quantification of this subsidy. Subsidies are complex, and this ratio provides only a simplistic assessment of the situation in any given utility.

For many, the cost of connecting to the piped-water network can be a significant financial hurdle. Comparing connection charges provides insight into the level of this hurdle, a point of particular concern when seeking to connect poorer sections of the community. The indicator provides cost of connection as an absolute level and as a proportion of national GNI per capita.

Billing customers and getting paid are two different things. The effectiveness of the collections process is measured by the amount of outstanding revenues at year-end, compared to the total billed revenue for the year, in day equivalents, and to the total amount collected as a percentage of the billed amount.

Financial Performance

Indicators

Indicator	Unit	Concept
24.1 Operating Cost Coverage	ratio	Total annual operational revenues/total annual operating costs
25.1 Debt Service Ratio	%	Cash income/debt service × 100

Source: Authors.

Discussion

The operating cost coverage ratio and the debt service ratio were selected from a much larger range of financial indicators (including leverage, liquidity, profitability, and efficiency ratios) because they help answer two important questions: Do revenues exceed operating costs? Does the utility's income enable it to service its debts?

Assets

Indicators

Indicator	Unit	Concept
27.1 Gross Fixed Assets—Water & Wastewater	US$/W and WW pop served	Total gross fixed water (W) and wastewater (WW) assets per water and wastewater populations served
27.2 Gross Fixed Assets—Water	US$/W pop served	Total gross fixed assets per population served, separately for water (W) and wastewater (WW)
27.3 Gross Fixed Assets—Wastewater	US$/WW pop served	

Source: Authors.
Note: pop = population.

Discussion

Gross fixed assets are defined to include work in progress.

The capital intensity of the utility is indicated by the gross fixed-asset value per capita served. Unfortunately, information about asset values is often limited. Until more emphasis is placed on this item, the values derived must be treated with caution.

No investment indicators are included. This reflects the difficulty of making meaningful comparisons at this high level between utilities with widely differing situations and investment needs. At a more detailed level, comparisons of unit costs for particular items of equipment can be very useful; this is beyond the scope of IBNET.

Affordability/Purchasing Power Parity

Gross national income can be converted from the local currency to U.S. dollars in two ways: by using the official exchange rate or by using purchasing power parity (PPP).

PPP takes account of what can be bought locally and should be considered for indicators of what customers pay. These indicators include the following:

Indicator	Unit
18.1 Average Revenue—W and WW	US$/m^3 water sold
18.2 Average Revenue—W and WW	US$/W conn/yr
18.3 Average Revenue—Water Only	US$/m^3 water sold
19.1 Total Revenues per Service Pop/GNI	% GNI per capita
19.2 Monthly Water Bill (for a household consuming 6m^3 of water per month through a household or shared yard tap but excluding the use of standpipes)	% GNI per capita
20.1 Residential Fixed Component of Tariff	US$/conn/yr
20.3 Residential Fixed Component of Tariff—Water	US$/conn/yr
20.4 Residential Fixed Component of Tariff—Wastewater	US$/conn/yr
22.1 Connection Charge—Water	US$/conn
22.2 Connection Charge—Water	% GNI per capita
22.3 Connection Charge—Sewerage	US$/conn
22.4 Connection Charge—Sewerage	% GNI per capita

Source: Authors.
Note: conn = connection, m^3 = cubic meter, W = water, WW = wastewater, yr = year.

References

Celine Nauges and Caroline van den Berg. 2008. "Economies of Density, Scale and Scope in the Water Supply and Sewerage Sector: A Study of Four Developing and Transition Economies." *Journal of Regulatory Economics* 34: 144–63.

Foster, Vivien. 2008. "Africa Infrastructure Country Diagnostic: Overhauling the Engine of Growth—Infrastructure in Africa." Executive Summary, World Bank, Washington, DC.

Global Water Intelligence. 2007. *Global Water Market 2008: Opportunities in Scarcity and Environmental Regulation.* Oxford, UK.

———. 2009. *Global Water Market 2010: Opportunities in Scarcity and Environmental Regulation.* Oxford, UK.

Komives, Kristin, Vivien Foster, Jonathan Halpern, and Quentin Wodon, with support from Roohi Abdullah. 2005. *Water, Electricity and the Poor: Who Benefits from Utility Subsidies?* Washington, DC: World Bank.

Milanovic, Branko, and Lire Ersado. 2008. "Reform and Inequality During the Transition: An Analysis Using Household Panel Data 1990–2005." Policy Research Working Paper 4780, World Bank, Washington, DC.

WHO (World Health Organization) and UNICEF (United Nations Children's Fund). 2000. *Global Water Supply and Sanitation Assessment 2000 Report.* Geneva: WHO; New York: United Nations.

———. 2010. *Progress on Sanitation and Drinking Water: 2010 update.* Geneva: WHO; New York: United Nations.

Index

Boxes, figures, and tables are indicated by *b*, *f*, and *t*, respectively.

A

ADERASA (Association of Water and Sanitation Regulatory Entities of the Americas), 3

affordability of water and sanitation services
as performance indicator, 142–143, 143*t*
trends in, 27–30, 28*t*, 29*f*, 29*t*

Africa. *See* Middle East and North Africa; Sub-Saharan Africa; specific countries

Africa Infrastructure Country Diagnostic studies, 28

Albania
data table, 47*t*
inauguration of benchmarking efforts in, 7

AMAC (Moldova Apa Canal), 6*b*

Apgar score for water and sanitation utilities, 30–34, 31*t*, 32*f*, 33*f*, 33*t*, 34*t*

Argentina, data table for, 48*t*

Armenia
data table, 49*t*
inauguration of benchmarking efforts in, 7
user-generated country report, 14*f*

Asia. *See* East Asia and Pacific; Europe and Central Asia; South Asia; specific countries

assets, as performance indicator, 142*t*

Association of Water and Sanitation Regulatory Entities of the Americas (ADERASA), 3

Atlas method, 141

Australia
data table, 50*t*
participation in IBNET, 15

B

Bangladesh
data table, 51*t*
inauguration of benchmarking efforts in, 7

Belarus
data table, 52*t*
inauguration of benchmarking efforts in, 7

benchmarking
defined, 2–3
IBNET benchmarking tool, 3
metric, 3
process, xi, 3
steps in, 14

Benin, data table for, 53*t*

Bhutan, data table for, 54*t*

billing and collections, 27*t*, 140–141*t*, 141–142

blockages, bursts, and breaks, 137–138, 137*t*

Bolivia, data table for, 55*t*

Bosnia and Herzegovina, data table for, 56*t*

Brazil
data table, 57*t*
economies of scale in, 34*b*
reform driven by benchmarking in, 5*b*

breaks, bursts, and blockages, 137–138, 137*t*

Bucknall, Julia, xii

Bulgaria
data table, 58*t*
technical assistance agreement with IBNET, 7

Burkina Faso, data table for, 59*t*

bursts, blockages, and breaks, 137–138, 137*t*

Burundi, data table for, 60*t*

business planning, 35–36, 36*f*

T

Tajikistan, data table for, 122*t*
Tanzania
    data table, 123*t*
    O&M costs in, 24
tariffs, 8, 10*f*, 43–44, 43*f*, 141
time-series data, development of, 4, 36
Togo, data table for, 124*t*
transparency, IBNET promoting, 11
trend analysis, 17
trends in water and sanitation sector
    status. *See* water and sanitation
    sector
Tunisia, data table for, 125*t*
Turkey, data table for, 126*t*

U

Uganda, data table for, 127*t*
Ukraine
    Chisinau Apa Canal compared to, 36,
        39, 40, 40*f*, 41*f*, 43, 44*f*, 45*f*, 46
    data table, 128*t*
    inauguration of benchmarking
        efforts in, 7
unaccounted-for water, 136–137
UNICEF-WHO Joint Monitoring
    Program 2008 MDG assessment, 6,
    21, 135
unit operational costs, 138–139, 138*t*
United Kingdom Department for
    International Development (DFID),
    xi, xii, 2, 7
United States, number of utilities in
    IBNET in, 8*t*
Uruguay, data table for, 129*t*
user-generated country reports, 13, 14*f*
Uzbekistan, data table for, 130*t*

V

Venezuela, República Bolivariana de, data
    table for, 131*t*
verification of data, 16
Vietnam
    data table, 132*t*
    economies of scale in, 34*b*

technical assistance agreement with
    IBNET, 7
voluntary nature of participation in
    IBNET, 3–4, 15

W

wastewater coverage, 19*t*, 23, 24
Water Action Plan, DFID, xii
Water and Sanitation Program–South
    Asia, 7
water and sanitation sector, 17–34
    affordability, 27–30, 28*t*, 29*f*, 29*t*
    Apgar score for, 30–34, 31*t*, 32*f*, 33*f*,
        33*t*, 34*t*
    collection period, 27*t*
    economies of scale in, 34*b*
    IBNET for. *See* International
        Benchmarking Network for Water
        and Sanitation Utilities
    increasing demands on, xi, 1
    international reporting on status of, 8
    NRW, 19–20, 20*t*, 21*f*, 21*t*, 22*f*
    operating cost coverage ratio, 11*f*,
        23–27, 23*t*, 24–26*f*, 25*t*, 26*t*
    performance measures for, 17–18
    size of utility, 32, 33*f*, 34*b*
    staff productivity, 20–23, 23*t*
    subsidies, 28–30, 29*t*, 141
    trend analysis, 17
    wastewater coverage, 19*t*, 23, 24
    water supply coverage, 18*t*
water consumption and production,
    135–136*t*, 136
Water Performance Assessment Start-Up
    Toolkit, 6*b*
water tariffs, 8, 10*f*, 43–44, 43*f*, 141
Western Europe. *See* Europe and Central
    Asia, and specific countries
World Bank
    IBNET administered by, xi, 2
    strategic involvement in water and
        sanitation sector, xii

Z

Zambia, data table for, 133*t*